GROWTH FACTORS

IN FOCUS

Titles published in the series:

*Antigen-presenting Cells
*Complement
Cytokines
DNA Replication
Enzyme Kinetics
Gene Structure and Transcription 2nd edition
Genetic Engineering
Growth Factors
*Immune Recognition
Intracellular Protein Degradation
*B Lymphocytes
*Lymphokines
Membrane Structure and Function
Molecular Basis of Inherited Disease 2nd edition
Molecular Genetic Ecology
Protein Biosynthesis
Protein Engineering
Protein Structure
Protein Targeting and Secretion
Regulation of Enzyme Activity
*The Thymus

*Published in association with the British Society for Immunology.

Series editors

David Rickwood
Department of Biology, University of Essex, Wivenhoe Park,
Colchester, Essex CO4 3SQ, UK

David Male
Institute of Psychiatry, De Crespigny Park, Denmark Hill,
London SE5 8AF, UK

GROWTH FACTORS

John K. Heath

Department of Biochemistry, University of Oxford,
South Parks Road, Oxford OX1 3QU

IRL PRESS
—at—
OXFORD UNIVERSITY PRESS

Oxford University Press, Walton Street, Oxford OX2 6DP

Oxford   New York   Toronto
Delhi   Bombay   Calcutta   Madras   Karachi
Kuala Lumpur   Singapore   Hong Kong   Tokyo
Nairobi   Dar es Salaam   Cape Town
Melbourne   Auckland   Madrid
and associated companies in
Berlin   Ibadan

Oxford is a trade mark of Oxford University Press

In Focus is a registered trade mark of the Chancellor, Masters, and Scholars
of the University of Oxford trading as Oxford University Press

Published in the United States
by Oxford University Press Inc., New York

© *Oxford University Press, 1993*

A catalogue record for this book is available from the British Library

Library of Congress Cataloging in Publication Data
Heath, John K.
Growth factors / John K. Heath. – 1st ed.
Includes bibliographical references and index.
1. Growth factors.   I. Title.   II. Series: In focus (Oxford,
England)
[DNLM 1. Growth Substances.   QU 100 H437g 1993]
QP552.G76H43   1993   599'.08761–dc20   93–7372
ISBN 0–19–963041–0 (pbk.)

Typeset by Footnote Graphics, Warminster, Wiltshire
Printed by Interprint Ltd, Malta

Preface

Growth factors, as the name implies, were originally discovered by virtue of their ability to promote cell multiplication in culture. The last few years has seen an explosion of discoveries in this area and growth factors are now implicated in many fields of modern biology including direct clinical applications as well as more fundamental scientific problems. This book is designed to provide an introduction to this increasingly significant class of intercellular signalling molecules.

The starting point of the text is the historical roots of growth factor research in the biology of mammalian cell proliferation. This leads into a more detailed consideration of the biochemistry of growth factors, their receptors and intracellular signalling pathways. This focus on the fundamentals is deliberate. The scope of growth factor biology is vast, increasing, and impossible to adequately cover in a simple monograph. It is therefore aimed to equip the interested reader with sufficient background knowledge to venture further into the specialized literature.

It is also difficult to write a book on growth factors without venturing into other fields. In particular, cancer biology, development biology, and neuroscience are beginning to have a profound impact on the understanding of growth factor function. This text must therefore be viewed as a status report. The definitive account is many years in the future.

I am very grateful to the many students, colleagues, and collaborators whose ideas and comments have been an essential stimulus to the composition of this book.

John K. Heath

Contents

Contents

4. Growth factors and the nuclear response

Abbreviations

aFGF	acidic fibroblast growth factor
bFGF	basic fibroblast growth factor
BMP	bone morphogenetic protein
BSA	bovine serum albumin
CNTF	ciliary neurotrophic factor
DAG	diacylglycerol
EGF	epidermal growth factor
EGF-R	epidermal growth factor receptor
FGF	fibroblast growth factor
FSH	follicle stimulating hormone
G1	phase following the mitotic phase of the cell cycle
G2	phase following the S phase of the cell cycle
GAP	GTPase activating protein
G-CSF	granulocyte colony stimulating factor
GM-CSF	granulocyte macrophage colony stimulating factor
HeLa	a carcinomatous cell type, named after Henrietta Lacks
IGF	insulin-like growth factor
IGF-BP	insulin-like growth factor binding protein
K-FGF/Hst	kaposi fibroblast growth factor
KGF	keratinocyte-derived growth factor
LAP	latency associated protein
LHRH	luteinizing-hormone releasing hormone
M-CSF	macrophage colony stimulating factor
MIS	mullerian inhibitory substance
M phase	mitotic phase of the cell cycle
MSV	moloney sarcoma virus
NGF	nerve growth factor
PDGF	platelet-derived growth factor
PIP2	phosphatidylinositol 4,5 bisphosphate
PKC	protein kinase C
PLC	phospholipase C
PTH	parathyroid hormone
S phase	synthesis phase of the cell cycle following the G1 phase
TGF	transforming growth factor

TRE	TPA response element
TRH	thyrotrophin releasing hormone
VEGF	vascular endothelial cell growth factor
VVGF	Vaccinia virus growth factor

1

An introduction to cell proliferation

1. Introduction

The ability to replicate and multiply is a fundamental and characteristic property of living organisms from the simplest single-celled bacterium to the most complex multicellular vertebrate such as ourselves. At the level of the individual cell, however, the biological demands are quite distinct. A multicellular organism is a community of cells whose multiplication must be matched to the overall requirements of the organism, and this demands that the behaviour of individuals cells must be subject to some form of coordinate control. In other words, the multiplication of cells within a multicellular organism is responsive to external signals generated by other cells.

In this chapter, the multiplication of mammalian cells is considered from a biological perspective. This definition of the behavioural characteristics of mammalian cell multiplication provides the basis for understanding the nature of the underlying mechanisms.

2. The cell cycle

The multiplication of any cell involves two essential phases—replication of the genetic material and equal partitioning of the replicated DNA into two daughter cells (mitosis). This process is quasi-repetitive in that, except in particular circumstances (described below), the processes are repeated in the daughter cells and in their daughter cells in turn. In addition, the two processes occur in a fixed order. The replication of DNA must proceed, and be completed, before it is partitioned into the cells of the next generation. There are some exceptions to this generalization. The most important, which occurs in all sexually reproducing organisms, is the generation of haploid gametes in which genetic information is distributed without prior replication (meiosis). In some specialized cells, DNA is replicated without a succeeding mitosis, leading to the production of individual cells with an increased DNA content. The processes involved in the

1

orderly completion of both phases in the life of a cell is accordingly called the cell cycle.

2.1 Measurement of the phases of the cell cycle

DNA replication and mitosis represent 'landmarks' in the cell cycle which can be readily observed and measured by the investigator. When cells undergo DNA replication they incorporate nucleotide precursors into their DNA and this can be detected by exposing the cells to chemically or radioactively labelled nucleotide precursors. As cells enter mitosis their chromosomes condense and become orientated on the mitotic spindle, which is followed shortly after by partitioning of the chromosomes and division of the cell into two daughters. The process of mitosis and its individual phases can be easily visualized under the microscope. In addition, when the cell divides, two daughter cells exist in place of the original parent. In the absence of cell death, a population of dividing cells increases in number. The practical significance of this is that the ability to observe and measure these landmark events provides the basis for defining the kinetic and regulatory properties of the cell cycle.

It is not possible to measure directly the duration of every phase in the cell cycle in a population of cells. Instead it is necessary to bring together a variety of approaches to build a composite picture of cell cycle characteristics of a population of cells undergoing multiplication. These approaches fall into three main categories, each of which provides certain types of information.

Firstly, it is possible to sample a population of cells at different times and determine experimentally either the total number of cells present or the fraction in S phase (the 'synthesis phase' of the cycle) or mitosis, or, by means of flow cytofluorimetry, determine the fraction of the population which have doubled their DNA content but have not yet divided. This sampling approach provides information on the overall properties of a population of cells, but little information on the duration of individual phases or the behaviour of individual cells within the population. Such a sampling approach can, however, become quite sophisticated when measurement of the two landmark events is combined, as, for example, in exposing cells to labelled nucleotide precursors for brief periods and then measuring the rate at which labelled cells reach mitosis.

The second approach involves the continuous observation of a population of dividing cells by time-lapse film recording. This permits detailed analysis of individual cell behaviour but is restricted to the measurement of mitosis or cell division, which are the only phases measurable by direct observation.

The third approach relies on the use of drugs which block progress through the cell cycle at either S phase (e.g. hydroxyurea) or mitosis (e.g. Colcemid). If such a drug-induced block to progress through the cell cycle is applied, as individual cells reach the point of the block further progress is arrested and a population of cells accumulates at a particular point in the cell cycle. This phenomenon can be used in two ways. Firstly it may be possible to measure the rate at which cells accumulate at the block, and thereby infer the rate of progress through the previous phases, or cells may be allowed to accumulate at the block for a fixed period, after which the block is released by removing the drug, and the time taken to reach the next observable landmark event measured. This type of experiment is especially useful in measuring the duration of periods of the cell cycle in which no easily measurable landmark event is occurring. The drawback of this type of approach is, of course, the possibility that the duration of cell cycle events being measured are influenced by the action of the drug.

2.2 Other features of the cell cycle

The combined application of the methods described reveals that the 'typical' eukaryotic cell cycle contains 'extra phases' in which the cells are neither undergoing DNA synthesis nor actively dividing. These have been termed G1, which is the period after the daughter cells have been formed but before they enter DNA synthesis, and G2, which is the phase between the completion of DNA synthesis and the onset of mitosis. In the G1 phase, cells have typically a diploid DNA content (2C) and in the G2 phase, they have twice the diploid DNA content (4C). It is now possible to consider the cell cycle, from the 'birth of a cell', as therefore comprising four phases: G1, the period of DNA replication (or S phase), G2, and mitosis (*Figure 1.1*). Each of these phases can be either directly measured or inferred and it is therefore possible to create a composite kinetic analysis of cell behaviour in each of the four cell cycle phases.

2.3 The kinetic properties of the phases of the cell cycle

Combining a large body of information obtained from different types of cells by different experimental means, allows us to make some general statements about the duration of the individual cell cycle phases.

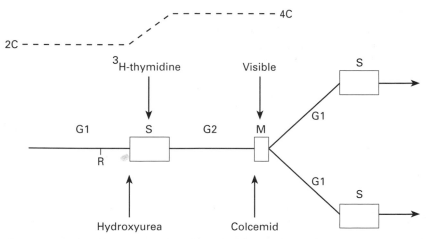

Figure 1.1.   The phases of the mammalian cell cycle. S phase can be detected by incorporation of labelled nucleotide precursors and mitosis can be visualized under the microscope. Progress through phases can be specifically blocked by drugs such as hydroxyurea (S phase) or Colcemid (M phase). Progress through S phase entails a doubling of DNA content from 2C to 4C. R = restriction point.

Firstly, mitosis is a relatively rapid affair, often occupying less than an hour, and which exhibits little variation in duration between cells with widely different overall rates of multiplication.

Secondly, S phase typically takes a matter of a few hours to complete (although there are striking exceptions, such as in the early embryos of certain species in which S phase occurs in 15–30 min). Likewise, G2 occupies a few hours in the mammalian cell's cycle except in special cases such as the unfertilized egg, in which G2 is prolonged indefinitely until it is completed suddenly upon fertilization. The duration of both S phase and G2 may vary slightly between individual cells within a population but this variation is, in the majority of cases, not correlated with the overall distribution of cell cycle times of the entire population, and, in particular, S and G2 have roughly the same duration in populations of cells whose rates of multiplication are quite dissimilar.

It is the duration of the G1 phase which appears to be the major determinant of the duration of the cell cycle in eukaryotic cells. The influence of the G1 phase arises, not necessarily from its absolute duration, but rather from the fact that it is a period which exhibits considerable variation between populations of cells multiplying at different rates. It is, in fact, a noteworthy feature that individual eukaryotic cells within a population rarely exhibit identical cell cycle times. In most cases the overall duration of the cell cycle in an apparently identical population of cells is variable. Furthermore, the degree of variability increases as the overall rate of population doubling decreases, and this arises because as the rate of population doubling slows, the duration of G1 extends and at the same time becomes more variable within a population, whilst the other phases remain fairly constant in duration. In the extreme case the G1 phase of the cell cycle can be considered to be 'indefinitely extended' and further progress into S phase, G2, and mitosis is halted.

Examination of the distribution of either the overall cell cycle or the duration of G1 reveals that the variability is not simply 'random' or Gaussian in nature but instead appears to become skewed towards longer and longer cell cycle times as the overall cell cycle becomes extended (*Figure 1.2*). Within this 'skewed distribution' there are cells whose G1 phase duration is little different from those of a more rapidly dividing population and other cells whose G1 phase has become 'infinitely extended' and, in effect, never divide again. As population doubling times slow further, it is this latter category of cells whose representation within the population increases. This allows us to define an additional parameter of population cell cycle kinetic: the 'growth fraction' of a population of cells, which is the proportion of cells within a population which will give rise to daughter progeny within a measurable time frame.

The variability of cell cycle times in eukaryotic cells has exercised a considerable degree of fascination for mathematicians, who have attempted to fit population cell cycle data to different types of mathematical distribution, and thereby deduce features of the mechanism underlying the variation. This has proved to be a difficult (and so far futile) exercise. What does emerge most clearly from consideration of the eukaryotic cell cycle is, however, the fact that variability,

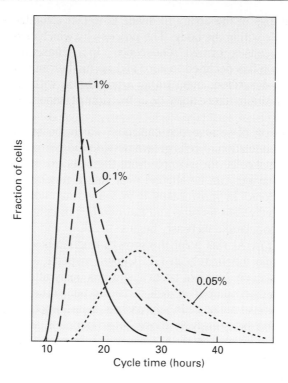

Figure 1.2 Distribution of cell-cycle times in populations of 3T3 cells grown in different concentrations of serum. As the overall rate of multiplication slows the distribution of cell-cycle times becomes more variable.

and hence control and regulatory processes, occurs predominantly in the G1 phase of the cell cycle. Any attempts at molecular explanations of cell cycle regulation and control must therefore focus, at least in part, on the biochemical processes of G1.

3. Senescence

So far the eukaryotic cell cycle has been considered to be an indefinitely repetitious, albeit variable, process. Common sense consideration of the human body, however, argues that this is an oversimplification. Whilst some cells in the adult, such as those of the skin or the blood, proliferate throughout most of our adult lives, others, such as those of the nervous system, cease dividing soon after birth and never divide again in the lifetime of the organism. Closer consideration of tissues such as the skin or the blood suggests, in fact that even within an actively dividing tissue, cells cease to divide and eventually die or are removed, and the two processes are normally in balance.

The eventual loss of the ability to divide is in fact a fundamental property of most types of cells within the body. The process by which cells lose the ability to produce progeny cells is termed 'senescence'. In a senescent population of cells the growth fraction has declined to zero and no further increase in population is possible. As just described, even within a population with cells actively undergoing cell proliferation, there may be a fraction of senescent cells that never divide.

The phenomenon of senescence can most easily be studied by longitudinal observation of a population of cells grown outside the normal habitat of the body, in culture. If normal cells, for example from the skin, are removed and grown in culture, their behaviour can be studied over many generations under relatively controlled conditions. In most cases it is found that normal cells divide and multiply quite rapidly, but as time and the number of generations increase, the overall rate of population increase begins to slow and eventually reaches a plateau (*Figure 1.3*). At this point the growth fraction has dropped to zero and all the cells remain indefinitely in G1. Closer inspection of this phenomenon shows that senescence does not occur instantaneously after the same period of chronological time or number of cell divisions but rather it is a progressive process. As a population of cells approaches senescence, cell cycle times increase and become progressively more heterogeneous and, at the same time the growth fraction progressively decreases (1). In other words the process of senescence, whilst seemingly an obligatory fate of all cells, exhibits considerable cell-to-cell variation.

The phenomenon of senescence also exhibits some unexpected properties when different types of cells are considered. Firstly, extensive studies of human fibroblasts derived from skin biopsies from different individuals show that the average number of cell doubling that occurs before the plateau phase is relatively constant from individual to individual, being, in this instance, approximately 50 generations (2). Secondly, the number of generations required for a population to reach the plateau phase is strongly influenced not by the species of origin, but by the exact type of cell in question. Whilst skin fibroblasts reach a plateau after approximately 50 generations, some cell types from the early embryo can plateau and senesce after only six to seven generations in culture. This means that cellular senescence appears to be a 'programmed' phenomenon whose kinetics are ultimately controlled by the identity of the type of cell.

4. Quiescence

Senescence is, as described above, an essentially irreversible phenomenon. A population of cells may, however, cease to proliferate, but retain the ability to divide further. This phenomenon is termed 'quiescence'. The state of quiescence is most usually brought about by manipulation of the environment of the cell, and how this occurs is described in detail in Section 7. The important point to note, however, is that quiescence is a reversible phenomenon. If cells are rendered

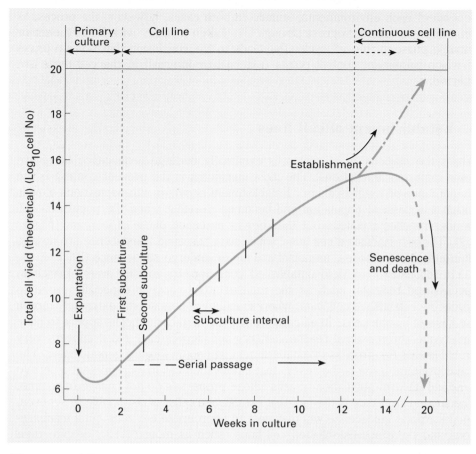

Figure 1.3 Evolution of a cell line. The vertical axis represents total cell growth (assuming no reduction at passage) on a log scale; the horizontal axis represents time in culture on a linear scale, for a hypothetical cell culture. Although a continuous cell line is depicted as arising at 12½ weeks it could, with different cultures, arise any time. Likewise, senescence may arise at any time, but for human diploid fibroblasts it is most likely between 30 and 50 cell doublings or 10–20 weeks depending on the doubling time. (Reproduced from ref. 2; with permission.)

quiescent by alteration or manipulation of their environment, they can be induced to re-enter the cell cycle and proliferate by restoration of the original conditions. Closer inspection of cells whose multiplication has been brought to a halt by quiescence reveals that they are halted, not randomly, but specifically in the G1 phase of the cell cycle, and the process of departure from quiescence and reinitiation of cell multiplication must necessarily involve events occurring in G1.

Senescence and quiescence can be considered as being analogous phenomena; they both involve cessation of multiplication and progress through the cell cycle. Whilst senescence is 'programmed' and essentially irreversible, quiescence is dependent upon environmental stimuli. In both cases, however, the process is initiated by failure to progress through G1. Taken together with G1 as the major variable phase in the cell cycle, this leads to the conclusion that G1 is a phase in which prominent control points (external or internal) in the cell cycle are exerted.

5. Establishment of cell lines

Whilst the majority of normal cells eventually undergo senescence there are some important exceptions. The most significant in the present context is the phenomenon of 'establishment'. Establishment occurs in sub-populations of cells within a senescent population and becomes manifest when the population cell number reaches a plateau but then begins a second phase of increase (*Figure 1.3*). These 'established cell lines' which have 'escaped' senescence are now to all intents and purposes, immortal and never undergo senescence (3).

The propensity to yield established cells appears to vary from species to species and rodents, such as the mouse or rat, appear especially prone to producing established cell lines, whereas establishment seems extremely rare in the case of human cells. In all cases, however, established cells appear to have undergone chromosomal rearrangements and duplications and it appears very probable that the process of establishment has some form of genetic basis. The major practical issue is, however, that established cell lines represent a permanent and uniform population of cells whose properties do not change over time. The second key issue is the fact that established cells, whilst unable to senesce, do still exhibit quiescence when subject to appropriate environmental manipulations and therefore established cell lines (such as mouse 3T3), despite their abnormalities, have become the most common cell type for biochemical and biological investigations into the control mechanisms of the cell cycle (3, 4).

6. Cell proliferation and the environment

It has already been mentioned that the multiplication of cells can, as manifested in the phenomenon of quiescence, be controlled by external environmental influences which appear to act on control points in the G1 phase of the cell cycle. The ability to control cell multiplication through manipulation of the environment is not only a powerful practical tool but it also provides a crucial entry point into the biochemical analysis of the control of mammalian cell proliferation. The philosophy behind this approach is that determination of the identity of the critical environmental determinants of cell proliferation, in particular those responsible for the phenomenon of quiescence, could lead to analysis of their action at the molecular level and thence into biochemical dissection of the intracellular control mechanisms themselves. The development of this approach occupies the remaining chapters of this book, but at this point it is valuable to consider further the biological nature of the response of the eukaryotic cell cycle to environmental factors.

Considerable effort has been expended in the past hundred years in trying, largely by empirical means, to find appropriate conditions to permit the propagation of mammalian cells in culture. For, whereas bacteria and fungi will, in most cases, proliferate in the presence of simple nutrient sources, the requirements of vertebrate (principally mammalian) cells are significantly more complex. In particular, mammalian cells appear to require environmental factors over and above simple chemical or macromolecular nutrients.

The identification of these extra requirements has been a lengthy and difficult process and, for many types of cells, is by no means complete. The experimental methods involved in resolving these issues comprise two contrasting approaches. The first and historically most common involves the attempt to recreate, as faithfully as possible, what is thought to represent the 'natural' environment of cells within the body. Having found conditions which permit cell proliferation, the next step is to begin biochemical fractionation with the objective of identifying the active components. The second, and more recent method, is to begin with a simple mixture of nutrients and to 'add back' ingredients until an optimal mixture for cell proliferation is obtained. As will be seen below these two approaches converge on the identification of a core set of molecules required for the survival and, most importantly proliferation of mammalian cells in culture.

7. The role of serum

The earliest successful attempts at growing mammalian cells in culture involved the use of clotted blood, either in the form of solid plasma clots (such as would occur at the site of wound injury) or serum (the soluble fraction of clotted blood). It was noted in these cases that serum (either solid, in the form of the clot, or liquid) was essential if cells were to proliferate, whereas viability could be maintained in the presence of simple buffered salt solutions fortified with nutrients. By a converging process of optimizing both the nutrient mixture components and the sources of serum, it proved possible, by the early 1950s, to grow a limited number of cell types in culture quite readily for prolonged periods of time (up to and including the plateau phase) under reproducible conditions. This technical achievement had two significant implications. It became possible to define experimentally in what way the presence of serum influenced progress through the cell cycle and thence to attempt to define the biochemical components of serum responsible.

The most significant investigations into the role of serum in the control of mammalian cell multiplication centred on the analysis of the established mouse cell line 3T3. Robert Holley and co-workers (5) noted that if 3T3 cells were placed in culture in a nutrient mixture supplemented with serum, they multiplied, then the population plateaued and entered quiescence (*Figure 1.4(a)*). The number of divisions that occurred before quiescence was related to the amount

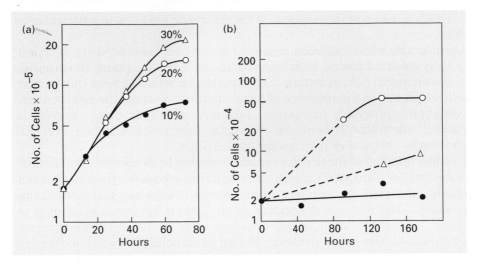

Figure 1.4.   Growth curves of 3T3 cells in (a) media containing 10, 20, and 30% calf serum and (b) medium depleted by growth of 3T3 cells until growth stopped: ●, without fresh calf serum; ○, with 10% fresh calf serum; △, depleted medium replaced after 20 h and 90 h with medium depleted by a three-day exposure to confluent 3T3 cells.

of serum present in the culture, thus cells exposed to 1% serum stop dividing after fewer generations than those exposed to 10% serum (6). Furthermore, cells that had become quiescent in the presence of 1% serum could not only be stimulated to redivide by the addition of fresh serum, but also the number of cell doublings that occurred was directly related to the amount of serum added— thus cells exposed to 1% serum, allowed to enter quiescence, and then re-exposed to fresh 10% serum achieved the same number of generations as those exposed to 10% serum from the beginning (*Figure 1.4(b)*).

The conclusions from these experiments are highly significant; the results imply that serum contains some substance(s) which induces cells to multiply and this substance is used up in the process of proliferation. When the 'proliferation-inducing' factor (or 'growth factor') is exhausted the cells cease to proliferate and become quiescent in G1, but progress through the cell cycle, and exit from G1 can be reinitiated by addition of fresh factor. In other words, in order to progress through G1 phase, 3T3 cells require the presence of external factors which can be supplied by serum.

This basic experimental observation was elaborated in two distinct ways. Firstly it could be used to show that the absolute amount of available growth factor present was the major determinant of subsequent cell multiplication rather than, for example, the proximity of neighbouring cells within the culture dish, and these factors were present in extremely small amounts. Secondly it was possible to determine whether other tissues, body fluids, or indeed the culture media of other cells contained, like serum, factors which induced exit from quiescence and cell proliferation. It rapidly became apparent that not only did a wide variety of tissues contain such growth factors but also that the requirements for such factors and their optimal sources varied widely from one cell type to another.

The general conclusion from this class of experiment is therefore that the control of progress through G1 is mediated by soluble factors, present in many tissues, and the exact requirements, both qualitative and quantitative, for these factors is dependent upon the exact cell type in question.

8. The role of no serum

The type of experiments described previously are complementary to the second approach to the identification of molecules required for the proliferation of cells in culture. In this approach, pioneered by Sato and co-workers, attempts were made to define the minimal requirements for cell proliferation in culture by adding back defined substances to culture medium containing only defined nutrients, vitamins, and salts. Whilst the components of such media are required for cell viability it is absolutely clear that these micronutrient components themselves are insufficient for cell proliferation (7). The fruits of this second approach were firstly an appreciation that the exogenous requirements for cell multiplication were complex but biochemically definable and secondly that cells require an optimal mixture of specific external macromolecular factors for proliferation (8). These latter factors include not only growth factors of the type defined from the study of serum-containing media, but additional factors (also supplied by serum) for cell attachment (such as the adhesion proteins fibronectin, laminin, and vitronectin) and cell nutrition (such as transferrin and plasma lipoproteins). Furthermore, the exact combination of defined factors required for the multiplication of cells in culture depends on the phenotype of the cells in question. The proliferation of neuronal cell types, for example, requires quite different mixtures of extracellular macromolecules from those of kidney cells (*Table 1.1*).

The 'serum' and 'no serum' approaches to the analysis of mammalian cell multiplication therefore arrive at the same conclusion: mammalian cells require specific and biochemically definable soluble factors in order to proliferate in culture. The identity of the factors required also depends on the phenotype of the target cell. These studies also provide the technical base from which to approach the purification and biochemical characterization of the active agents.

9. The restriction point

Before going on to consider the biochemistry of individual growth factors in detail (Chapter 2), it is worthwhile exploring some implications of the recognition that serum contains specific factors that control progress through G1, and in particular, the biological features of the regulatory events underlying G1 control of cell cycle progress. This is possible because the use of serum provides a means of

Table 1.1 Media additives for serum-free culture of established cell lines from different sources (after Barnes and Sato 1980)

Cell line	Source	Requirements
GH_3	Rat pituitary	Insulin (5 μg ml^{-1}), transferrin (5 μg ml^{-1}), FGF (1 ng ml^{-1}), TRH (1 ng ml^{-1}), PTH (0.5 ng ml^{-1}), somatomedin C (1 ng ml^{-1}).
HeLa	Human cervical carcinoma	Insulin (5 μg ml^{-1}), transferrin (5 μg ml^{-1}), EGF (10 ng ml^{-1}), FGF (50 ng ml^{-1}), hydrocortisone (50 nM).
MDCK	Dog kidney	Insulin (5 μg ml^{-1}), transferrin (5 μg ml^{-1}), hydrocortisone (50 nM), T3 (0.005 nM), prostaglandin E_1 (25 ng ml^{-1}). (Supports primary cultures of normal kidneys.)
B104	Rat neuroblastoma	Insulin (5 μg ml^{-1}), transferrin (100 μg ml^{-1}), CIg (10 μg ml^{-1}), progesterone (20 nM), putrescine (100 μM). (Supports primary cultures of nervous tissue and neuroblastomas from many species and rat phaeochromocytoma.)
M2R	Cloudman mouse melanoma	Insulin (5 μg ml^{-1}), transferrin (5 μg ml^{-1}), testosterone (10 nM), FSH (0.4 μg ml^{-1}), LHRH (10 ng ml^{-1}), NGF (3 ng ml^{-1}). (Supports other melanomas and melanomas in primary culture.)
C_6	Rat glioma	Insulin (2 μg ml^{-1}), transferrin (5 μg ml^{-1}), FGF (20 ng ml^{-1}), gimmel factor (5 μg ml^{-1}).
MCF-7	Human mammary carcinoma	Insulin (0.1 μg ml^{-1}), transferrin (25 μg ml^{-1}), EGF (100 ng ml^{-1}), CIg (7.5 μg ml^{-1}), prostaglandin $F_2\alpha$ (100 ng ml^{-1}), α-1 spreading protein (1 μg ml^{-1}). (Medium also supports growth of BT20 human mammary cells.)
RF-1	Normal rat ovarian cells	Insulin (2 μg ml^{-1}), transferrin (5 μg ml^{-1}), hydrocortisone (10 nM), T3 (0.3 nM), CIg (8 μg ml^{-1}). (Medium supports normal granulosa cells in primary culture.)
F9	Mouse teratocarcinoma	Insulin (1 μg ml^{-1}), transferrin (5 μg ml^{-1}), CIg (5 μg ml^{-1}).
TM4	Normal mouse testicular cells	Insulin (5 μg ml^{-1}), transferrin (5 μg ml^{-1}), EGF (3 ng ml^{-1}), T3 (0.5 nM), FSH (0.5 μg ml^{-1}), growth hormone (100 ng ml^{-1}), somatomedin C (1 ng ml^{-1}) retinoic acid (50 ng ml^{-1}).
T84	Human colonic adenocarcinoma	Insulin (2 μg ml^{-1}), transferrin (2 μg ml^{-1}), EGF (1 ng ml^{-1}), hydrocortisone (50 nM), glucagon (0.2 μg ml^{-1}), ascorbic acid (10 μg ml^{-1}).
SV40-3T3		Insulin (0.25 μg ml^{-1}), transferrin (0.5 μg ml^{-1}), CIg (7.5 μg ml^{-1}), fatty-acid-free BSA (1 mg ml^{-1}), linoleic acid (5 μg ml^{-1}).

regulating entry and exit from the cell cycle in a controlled fashion and thereby probing in more detail the underlying nature of the regulatory processes involved. These studies reveal that progress through G1 appears to involve passage through a major gateway or control point in the exit from the quiescent state.

If cells are rendered quiescent by deprivation of serum they can, as described previously, be induced to re-initiate progress through the cell cycle by addition of fresh serum as a source of growth factors. A question that arises at this point is whether these serum factors are required to be present throughout the duration of a whole cell cycle, or are only required for some specific stage or stages. If the latter were the case it would pinpoint phases of the cell cycle which were responsive to external signals and thus identify periods of importance for biochemical investigation.

This issue can be addressed by inducing cells to exit quiescence by addition of fresh serum, and then withdrawing the serum at various intervals afterwards. If serum is required throughout the cell cycle one would expect that cells should cease further progress at the stage that had been reached at the point of withdrawal, whereas if serum-controlled events occur at a specific point then one would expect that cells would be able to complete the cycle in the absence of serum once the critical point had been passed. The results of these experiments are clear (9): cells only require exposure to serum for a period in the G1 phase of the cell cycle, and will complete a cycle if serum is withdrawn after a period corresponding to a few hours before the onset of S phase. This period in G1, mapped by the loss of serum requirement for further progress, is called the restriction point. That the restriction point, thus defined, has some mechanistic relevance, is supported by the additional finding that it corresponds approximately to a period when cell cycle progress is especially sensitive to inhibition of protein and RNA synthesis (10). Taken together, these findings suggest that the role of growth factors is to initiate a process or processes that are cell autonomous in nature and which culminate, after many hours, in mitosis and cell division. These processes are initiated at the restriction point which therefore represents a major point of control in the cell cycle. The variability in cell cycle kinetics and the molecular basis of senescence and quiescence all lie in the biochemical identity of events that lead up to passage 'through' the restriction point, which are in turn initiated by the interaction of growth factors with the quiescent cell.

10. Further reading

Barnes,D. and Sato,G. (1980). Serum free culture: a unifying approach. *Cell* **22**, 649–55.
Gospodarawicz,D. and Moran,J. (1976). Growth factors in mammalian cell cycle. *Ann. Rev. Biochem.* **45**, 531–58.
Holley,R. (1975). Control of growth of mammalian cells in culture. *Nature* **258**, 487–90.
Pardee,A., Dubrow,R., Hamlin,J., and Kletzian,R. (1978). Animal cell cycle. *Ann. Rev. Biochem.* **47**, 715–50.
Prescott,D.M. (1976). The cell cycle and the control of reproduction. *Adv. Genet.* **18**, 99–177.

11. References

11.1 Cellular senescence

1. Absher,M., Absher,R., and Barnes,W. (1974). Genealogies of clones of diploid fibroblasts: cinematographic observations of cell division patterns in relation to population age. *Exp. Cell Res.* **88**, 95–104.
2. Hayflick,L. and Moorhead,P. (1961). The serial cultivation of human diploid cells. *Exp. Cell Res.* **25**, 585–621.

11.2 Establishment and crisis

3. Todaro,G. and Green,H. (1963). Quantitative studies of the growth of mouse embryo cells in culture and their development into established cell lines. *J. Cell Biol.* **17**, 299–313.
4. Brooks,R.F., Richmond,F.N., Riddle,P.N., and Richmond,K.M. (1984). Apparent heterogeneity in the response of quiescent swiss 3T3 cells to serum growth factors: implications for the transition probability model and parallels with 'cellular senescence' and 'competence'. *J. Cell Physiol.* **121**, 341–50.

11.3 The role of serum

5. Holley,R.W. and Kiernan,J.A. (1968). 'Contact inhibition' of cell division in 3T3 cells. *Proc. Nat. Acad. Sci. USA* **60**, 300–4.
6. Paul,D., Lipton,A., and Klinger,I. (1971). Serum factor requirements of normal and simian virus 40 transformed 3T3 mouse fibroblasts. *Proc. Nat. Acad. Sci. USA* **68**, 645–8.

11.4 Serum-free medium: nutrient optimization

7. McKeehan,W.L., McKeehan,K.A., and Calkins,D. (1981). Extracellular regulation of fibroblast multiplication. Quantitative differences in nutrient and serum factor requirements for multiplication of normal and SV40 virus-transformed human lung cells. *J. Biol. Chem.* **256**, 2973–81.

11.5 Serum-free medium: the role of growth factors

8. Hayashi,H. and Sato,G. (1976). Replacement of serum by hormones permits growth of cells in defined medium. *Nature* **259**, 132–4.

11.6 The restriction point

9. Pardee,A.B. (1974). A restriction point for control of normal animal cell proliferation. *Proc. Nat. Acad. Sci. USA* **71**, 1286–90.
10. Rossow,P., Riddle,V., and Pardee,A. (1979). Synthesis of labile, serum-dependent protein in early G1 controls animal cell growth. *Proc. Nat. Acad. Sci. USA* **76**, 4446–50.

2

Growth factors

---

1. Introduction

In Chapter 1 the multiplication of many types of eukaryotic cells was revealed to be dependent upon the presence of exogenous polypeptide signalling molecules—growth factors which, after interacting with the cell, set in motion a chain of events leading to the onset of DNA synthesis and mitosis. This chapter examines the origins, structure, and biological activities of growth factors.

2. General properties of growth factors

Taken together growth factors represent a relatively large group of polypeptides which share the common property of inducing cell multiplication both *in vivo* and *in vitro*. As will be discussed, growth factors do not, as a whole, share any common structural features, although they can be grouped into 'families' of molecules with related sequences and structures. All growth factors exert their biological effects on cell multiplication at very low concentrations (typically 10^{-9}–10^{-11} M) and this is reflected by the fact that, with a few exceptions, growth factors are typically present in natural sources at extremely low concentrations.

Growth factors differ from classical endocrine hormones such as insulin or growth hormone in two important ways. Firstly, endocrine hormones are typically synthesized in specialized glands (such as the pancreas, in the case of insulin, or the pituitary, in the case of growth hormone) whereas growth factors are often synthesized in multiple types of cells and tissues. Secondly, classical endocrine hormones are released into body fluids at the site of synthesis and are carried to their target tissue in the bloodstream. In other words, the target tissue is often physically remote from the site of synthesis. A hallmark of growth factors is that, in most instances, they act locally within the tissues in which they are synthesized. In most cases, therefore, growth factors cannot be detected in the bloodstream or other body fluids. This local action of growth factors has led to the concepts of 'paracrine' and 'autocrine' action. Paracrine action occurs when a growth factor is secreted by a cell and interacts with responsive cells in the immediate vicinity and autocrine action occurs when a growth factor is expressed by, and acts upon, the same cell (*Figure 2.1*). The principal value of the concepts of autocrine and paracrine action is that they draw attention to the distinction in modes of delivery of growth factors to responding target cells. It is important to note, however, that growth factors and classical endocrine hormones have many properties in common. In particular, certain growth factors may have analogous effects to endocrine hormones on specific target cells and conversely certain classical endocrine hormones and neurotransmitters can exhibit mitogenic or growth-promoting effects on others. The biological actions of growth factors are not, therefore, necessarily restricted to the induction of cell multiplication. Thus whilst the mode of growth factor dissemination may be distinctive, the ability to induce DNA synthesis and cell division is a property shared with other agents better known in other contexts.

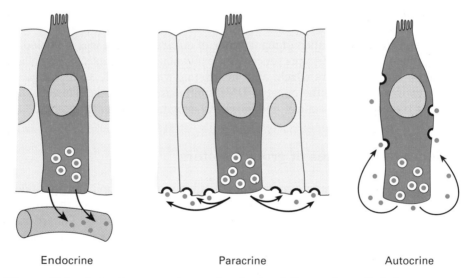

Endocrine   Paracrine   Autocrine

Figure 2.1   Autocrine, paracrine, and endocrine delivery of growth factors.

The fact that a relatively large number of different molecular species with mitogenic actions exists points to another important feature of growth factor action. This diversity of growth factor species reflects the fact that they exhibit cell-type specificity in their actions. The ability to induce cell multiplication is therefore a function of both the individual growth factor species and the phenotype of the responding cell. The 'host range' of growth factors is highly varied. Some growth factors exhibit a very restricted range of responsive target cell types, whereas other growth factors can induce the proliferation of many different (but not all) types of cells. On the other hand, individual cell types can also be characterized by the range of growth factors to which they are responsive. Some cell types (such as 3T3 cells) are able to respond to a number of different growth factors whereas others (such as erythroblasts) are highly restricted in the agents to which they respond.

What emerges from these considerations is the notion that growth factors are locally acting signalling polypeptides with diverse biological actions (including the ability to induce cell multiplication) which are dependent upon the identity of the responding cell type.

3. Growth factor families

3.1 The platelet-derived growth factor family

The identification of platelet-derived growth factor (PDGF) arose directly from the analysis of the role of serum in the induction of cell multiplication described in Chapter 1. It was noted that a number of cell types, notably 3T3 fibroblast cells and smooth muscle cells (1, 2), would proliferate in medium supplemented with serum (i.e. the soluble fraction of clotted blood) but their rate of proliferation was markedly reduced when the cells were cultured in plasma (i.e. the non-cellular fraction of whole blood). This finding suggested that a major growth factor (or factors) for these cells was released as a result of the process of blood clotting. The mitogenic activity of serum could, in addition, be restored by supplementing plasma with an extract of platelets (which break down in the process of clotting) (*Figure 2.2*). This finding strongly indicated that platelets contained the major mitogenic activity in serum for 3T3 and smooth muscle cells; this was termed platelet-derived growth factor (PDGF) (2).

The purification of PDGF from platelets proved a formidable undertaking (3) because, as it emerged, PDGF is not only present in platelet extracts at very low concentrations, but is also biologically active at very low concentrations. In other words a tiny amount of purified protein has a significant amount of biological activity. Despite these difficulties a number of groups were able to purify PDGF to homogeneity from human and porcine platelets. PDGF proved to be a dimeric molecule comprising two polypeptide chains (of about 12 kDa each) held together by disulphide bonds. Sequence analysis of PDGF from human platelets revealed that it is a heterodimer of two closely related polypeptide chains termed PDGF-A and PDGF-B. Porcine platelet-derived PDGF was, however, found to be a homodimer of two PDGF-B chains. PDGF was also purified from the culture medium of a human osteosarcoma cell line and this molecule proved to be a homodimer of two PDGF-A chains. PDGF therefore exists in three forms, the homodimeric PDGF-BB and -AA and the heterodimeric PDGF-AB (*Figure 2.3*).

The isolation of cDNAs of PDGF-A and -B (4, 5) permitted analysis of the tissue-specific expression of PDGF and it is now clear that PDGF is by no means restricted to platelets and megakaryocytes. PDGF has been found in a wide variety of tissues including the placenta, macrophages, smooth muscle cells from newborn animals, the central nervous system, and endothelial cells. PDGF was

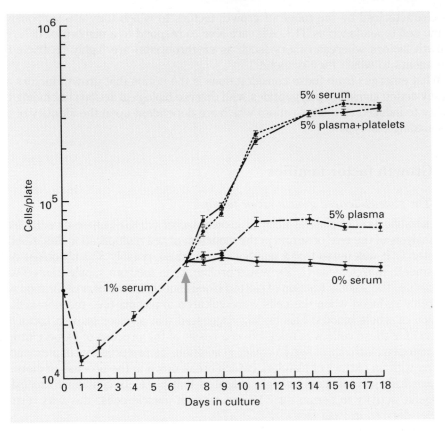

Figure 2.2. Growth of smooth muscle cells: the role of platelet-derived growth factors. Smooth muscle cells were cultured in the absence of serum (0% serum), 5% plasma, 5% serum, or 5% plasma to which an extract of platelets had been added. The growth-promoting effects of serum can be reproduced by adding platelet extract to plasma.

also found to be expressed by a wide variety of tumour cells in culture. Furthermore, PDGF-A and PDGF-B exhibit different patterns of expression; PDGF-B, for example, is predominantly expressed in the placenta and various adult tissues, whereas PDGF-A expression is prominent in the brain and in fetal tissues. This differential pattern of expression of the two PDGF chains suggests that each form of the PDGF molecule may have a subtly different biological activity which, as will be seen later, does indeed prove to be the case.

Analysis of vascular endothelial cell proliferation led to the discovery of vascular endothelial cell growth factor (VEGF) (16) which proves to be related to PDGF in sequence (*Figure 2.4*). Like PDGF, VEGF is composed of a dimer of two polypeptide chains. Despite its similarity in structure, VEGF has quite distinct biological activities and patterns of expression. PDGF is not therefore a

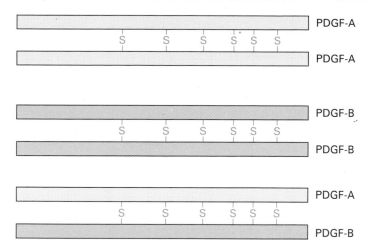

Figure 2.3.   The structure of platelet-derived growth factor (PDGF). PDGF occurs in three forms created by homo- or heterodimers of two distinct PDGF chains, A and B, linked by six disulphide bonds.

single entity, but a family of closely related molecules with differing biological activities and patterns of expression. In fact the existence of families of closely related molecules proves to be a theme which runs throughout the consideration of growth factor structure and function.

3.2 The epidermal growth factor family

3.2.1 EGF/urogastrone

Epidermal growth factor (EGF) was the first growth factor to be obtained in pure form and was discovered in a search for neurotrophic agents. Injection of extracts of mouse submaxillary gland into newborn mice was found to result in accelerated maturation of various epithelia, leading to premature eyelid opening and incisor eruption. Male submaxillary glands proved to be an extremely rich source of EGF and, as a result, purification of the molecule proved to be more straightforward than is the case for most growth factors. EGF isolated from mouse submaxillary glands is a 6 kDa polypeptide with mitogenic actions on a wide variety of epithelial cells (as well as 3T3 fibroblasts) in culture (7, 8).

Independent studies on gastric acid secretion led to the purification of a molecule, urogastrone, from the urine of pregnant women. This was initially characterized as an agent which inhibited gastric acid secretion. Upon sequence analysis, however, urogastrone proved to be closely related to mouse EGF and is, in fact, the human homologue of mouse EGF (9), exhibiting identical biological activities both *in vitro* and *in vivo*. It is noteworthy that the discovery of both human and mouse EGF arose from investigations that were not directly related to the issue of the control of cell proliferation, a finding which underscores the

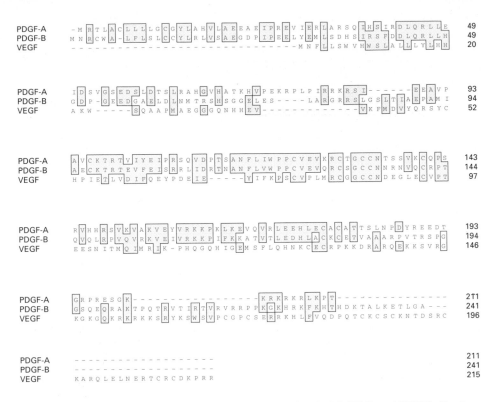

Figure 2.4.   Amino acid sequence alignment of PDGF-A, PDGF-B and VEGF. Regions of identity are boxed. Note the conservation of cysteine residues.

fact that many growth factors have biological activities other than the promotion of cell multiplication.

Molecular cloning of human and mouse EGF genes (10) revealed that the EGF polypeptide is part of a much larger precursor protein which is a transmembrane protein of apparent molecular mass 128 kDa. The mature EGF protein appears to be cleaved from the precursor by proteases present in the submaxillary gland. In addition the EGF precursor was also found to contain a number of repeats of amino acid sequences closely related to EGF. Studies on EGF expression also showed that, in contrast to many other growth factors, EGF expression was confined to the submaxillary gland, the kidney, and (in humans) Brunners gland in the gut.

These findings raise three interesting questions. Firstly, is EGF biologically active in the precursor form? Secondly, what is the physiological significance of the restricted expression of EGF? Finally, are other EGF-related sequences also capable of exhibiting growth factor activity? A partial solution to these issues became apparent upon the discovery of other growth factors related to EGF.

3.2.2 Transforming growth factor α

As mentioned previously, EGF is a potent growth factor for a variety of fibro-blast cell lines in culture. It was noticed, however, that infection of these lines by tumorigenic retroviruses such as moloney sarcoma virus (MSV) resulted in cells that were both able to form tumours in host animals and capable of sustained proliferation in the absence of EGF. One possible explanation of this phenom-enon was that infection by MSV resulted in the cells expressing their own 'EGF' in an autocrine (see *Figure 2.1*) mode thus becoming independent of exogenous EGF for multiplication. An EGF-like bioactivity was found in media conditioned by MSV-infected cells which proved, upon purification, to be a molecule which was related to, but not identical with, mouse EGF. This factor was called transforming growth factor (TGF) and proved to have almost identical biological actions to EGF and urogastrone.

TGFα is a 6 kDa polypeptide whose sequence is highly conserved between man and mouse and which shares about 33–44% homology with EGF (*Figure 2.5*). Like EGF, TGFα is made as a much larger 1200-residue amino acid trans-membrane precursor and cleaved from the precursor by specific proteases (11).

```
EGFs      1           2         3            4    5              6
Hu   NSDSE C PLSHDGY C LHD G V C   MYIEALDKYA   C N C VV GY I G ER C QYRDLKWWELR
Mo   NSYPG C PSSYDGY C LNG G V C   MHIESLDSYT   C N C VI GY S G DR C QTRDLRWWELR
Rt   NSNTG C PPSYDGY C LNG G V C   MYVESVDRYV   C N C VI GY I G ER C QHRDLR
Gp   QDAPG C PPSHDGY C LHG G V C   MHIESLNTYA   C N C VI GY V G ER C EHQDLDLWE
Pg   NSYSE C PPSHDGY C LHG G V C   MYIEAVDSYA   C N C VF GY V G ER C QHRDLKWWEL
Ho   NSYQE C SQSYDGY C LHG G V C   VYLVQVDTHA   C N C VV GY V G ER C QHQDLRWWELR
Rb   NSFPG C PPSHDGY C LHG G V C   MYIEAVDNYA   C N C VV GY V G ER C QHRDLKWWELR


TGF αs    1           2         3            4    5              6
Hu  VVSHFND C PDSHTQF C FH G T C   RFLVQEDKPA   C V C HS GY V G AR C EHADLLA
Rt  VVSHFNK C PDSHTQY C FH G T C   RFLVQEEKPA   C V C HS GY V G AR C EHADLLA
Bo  VVSHPND C PDSHSQF C FH G T C   RFLVQEEKPA   C V C HS GY V G AR C EHADLLA


Others       1           2          3              4    5              6
VGF   AIRL C GPEGDGY C LH G D C   IHARDIDGMY     C R C SH GY T G IR C QHVVLVDYQRS
MVGF       C NDDYKNY C LNN G T C  FTVALNNVSLNPF C A C HI NY V G SR C QFINLITIK
SFGF       C NHDYENY C LNN G T C  FTIALDNVSITPF C V C RI NY E G SR C QFINLVTY

AR   KKKNP C NAEFQNF C IH G E C   KYIEHLEAVT    C K C QQ EY F G ER C GEK
HB   KKRDP C LRKYKDF C IH G E C   KYVKELRAPS    C I C HP GY H G ER C HGLSL
```

Figure 2.5. Sequences of members of the EGF family which are known or are believed to activate the EGF receptor. Conserved residues are boxed; note the conser-vation of cysteine residues. Hu, human; Mo, mouse; Rt, rat; Gp, guinea pig; Pg, pig; Ho, horse; Rb, rabbit; Bo, bovine; VGF, Vaccinia virus growth factor; MVGF, myxoma virus growth factor; SFGF, Shope fibroma virus growth factor; AR, amphiregulin; HB, heparin-binding EGF-like growth factor.

Unlike EGF, however, TGFα expression is widespread in different tissues of both the fetus and adult as well as being expressed by a wide variety of tumour cells. The precursor form of TGFα is also able to induce a mitogenic response if cells expressing the molecule are in physical contact with responsive target cells.

The discovery of TGFα demonstrates three things. Firstly, that EGF is not a unique entity but, rather, is one of a larger family of molecules with related structures and biological activity. Secondly, EGF-like molecules are present in a wide variety of tissues apart from the submaxillary gland, and thirdly, the existence of a biologically active transmembrane precursor provides an example of how the action of growth factors could be highly restricted to autocrine and paracrine modes of action *in vivo*.

3.2.3 Other members of the EGF family

It is now apparent that the EGF family of growth factors is, in fact, sizeable. More detailed analysis of the expression of EGF-like bioactivities from tumour cell lines uncovered another EGF-like molecule—amphiregulin (12). Amphiregulin is a 96-residue amino acid secreted polypeptide with homology to the EGF prototype (*Figure 2.5*) which is expressed by a variety of tumour cell lines in culture whose expression pattern remains to be determined. Additional EGF-like molecules have also been found encoded by pox viruses and one such molecule, vaccinia virus growth factor (VVGF), has also been purified to homogeneity and found to exhibit very similar biological activities to EGF and TGFα (*Figure 2.5*). The existence of virally encoded EGF-like growth factors raises the possibility that pox viruses may exploit the biological actions of host-derived EGFs in the process of viral infection and reproduction.

As the sequence of many diverse proteins became available an additional feature of the EGF family emerged. Many different proteins, with biological functions in most cases unrelated to cell multiplication, contained sequences related to that of EGF. In at least a few (and probably most) cases, it is clear that these EGF-like sequences are not biologically active in terms of their ability to induce a cell multiplication. Instead, they represent a structural motif (or module) which can be used for a diversity of purposes in different contexts. Determination of the three-dimensional structure of EGF (13), TGFα, and the EGF 'module' from factor IX (the blood clotting protein) reveals that amino acids conserved amongst the EGF family of molecules (especially the characteristic cysteine motifs) are involved in the formation of a canonical structure and that biological function is determined by the identity of specific residues within the overall structure (*Figure 2.6*). This unexpected finding of a diversity of EGF-like molecules suggests that, at least in this case, growth factor activity may have evolved from molecules with entirely different biological function.

3.3 The fibroblast growth factor family
3.3.1 Acidic and basic FGF

Fibroblast growth factor (FGF) was discovered, as its name suggests, from its ability to induce the proliferation of fibroblast cells, such as 3T3, *in vitro*. FGF

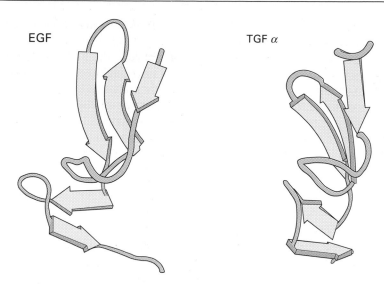

Figure 2.6   Three-dimensional structure of EGF and TGFα. Note the similarity in tertiary structure despite differences in amino acid sequence. (From ref 13; with permission.)

was first identified as a mitogenic activity present in commercial pituitary hormone preparations and, after a great deal of difficulty, was finally purified from bovine brain. A key feature of this purification was exploitation of the fact that FGF binds avidly to immobilized heparin (a polysulphated glycosaminoglycan found in many tissues) and affinity for heparin was, until recently, considered to be diagnostic for FGF (14). The biological significance of this characteristic feature of the FGFs is that they interact avidly with the heparin-containing proteoglycans present in many tissues and are, therefore, in effect, immobilized *in situ* and not disseminated far from their site of synthesis. This feature therefore represents another mechanism by which the biological action of growth factors may be constrained to a strictly local or paracrine context.

Initial purification of FGF from bovine brain (and a variety of other sources) quickly revealed that FGF was not one, but two, closely related molecules, with very similar biochemical and biological properties. On the basis of their pI and affinity for heparin they were termed basic FGF (bFGF) and acidic FGF (aFGF), and are powerful mitogens for a wide variety of cell types, including fibroblasts, neuroectodermal cells, and, most notably, capillary and large-vessel endothelial cells. Expression studies revealed that the expression of bFGF is extremely widespread in both adult and fetal tissues as well being found in a large number of both normal and tumorigenic cell lines (including large-vessel endothelial cells). The expression of aFGF is, by contrast, restricted principally to cells of the central and peripheral nervous system. An additional curious feature of both aFGF and bFGF is that they lack a characteristic secretory signal sequence at their amino termini (15, 16) and, in fact, neither are usually secreted from cells.

This would tend to indicate that they are not normally made available to responsive cells unless released by mechanisms brought about by cell damage.

Thus, as in the case of the PDGF and EGF families, the FGF family comprises apparently similar molecules (in terms of structure and biological function) which exhibit distinct patterns of expression in different body tissues and, in the case of bFGF, a growth factor found in nearly every tissue of the adult body. In this respect bFGF represents an extreme contrast to a typical endocrine hormone, being essentially local in its action and yet widespread in its activity.

3.3.2 FGF family members

For some time it was thought aFGF and bFGF were the sole representatives of this particular class of growth factors but further work, in (again) tangentially related areas, has uncovered five additional members of the FGF family of growth factors. The first of these, *int*-2, was identified (17) as a gene located close to a frequent insertion site of retroviruses in virally induced mammary tumours. Cloning and sequence of the *int*-2 gene revealed a molecule with significant sequence similarity to the aFGF and bFGF prototype genes (*Figure 2.7*). Further studies of genes that were capable of inducing growth-factor-independent proliferation of 3T3 cells led to the discovery of K-FGF/Hst (FGF-4) (18) and FGF-5 (19). Molecular cloning of genes related to FGF-4 led to the discovery of FGF-6 (20), a molecule sharing 80% identity in amino acid sequence with FGF-4. Finally an additional FGF-like growth factor, KGF (FGF-7) (21) was discovered as a major growth factor for keratinocyte-derived cell lines in culture. An important feature of these other FGF family members is that they all contain functional secretory signal sequences, and are therefore readily exported from cells. This suggests that at least one rationale for the existence of multiple related family members may, in this case, be to provide a means of controlling the dissemination and export of these molecules *in vivo*, by means of selective expression of individual family members. An additional (although as yet unproven) possibility is that, unlike the EGF family, each FGF family member may exhibit subtly different biological activities *in vitro* and *in vivo*.

Sequence alignment of the FGF family (*Figure 2.7*) shows that the overall similarity of the entire family is quite low, but that certain key residues are conserved in all cases and these residues (as might be anticipated from the analysis of the EGF family) appear to correspond to key sites in the three-dimensional structure of bFGF. The FGF family accordingly represents another example in which biological specificity is achieved through modifications of a common structural 'backbone'.

3.4 The insulin-like growth factors

The insulin-like growth factors (IGFs) represent a situation in which the distinction between growth factors and endocrine hormones becomes blurred. The existence of IGFs was first predicted in the form of the 'somatomedin hypothesis' which argued that the effects of pituitary growth hormone on skeletal

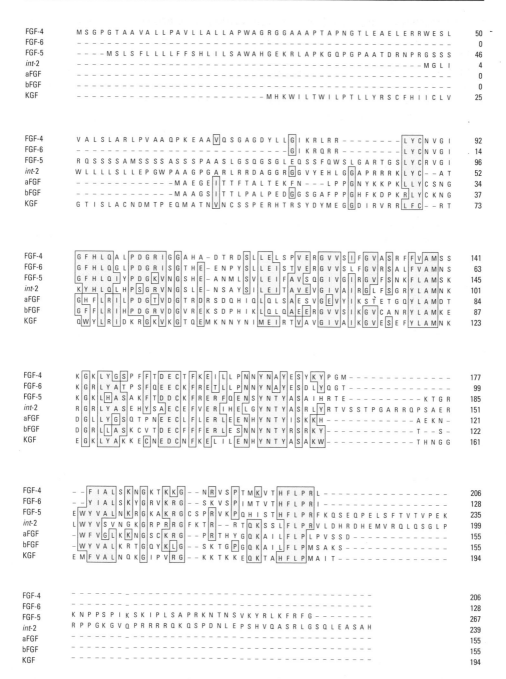

Figure 2.7 Amino acid sequence alignment of the FGF family of growth factors. Regions of amino acid identity are boxed. Note the sequence divergence between family members in the N-terminal region corresponding to the presence or absence of a secretory signal sequence.

growth were mediated by means of an intermediate class of bioactive peptides, the somatomedins (22). Following another heroic purification experiment (involving literally tons of starting material) two purified molecules were obtained, from plasma extracts, which were able to exert growth-hormone like effects on cartilage explants *in vitro*. Sequence analysis of these polypeptides revealed that both were related to proinsulin in sequence (*Figure 2.8*) (23), leading to their being designated insulin-like growth factors 1 and 2 (IGF-I and IGF-II). Both IGF-I and IGF-II are made as secreted prohormones (9 kDa and 14 kDa respectively) and require proteolytic cleavage to achieve their 6 kDa form. Although similar in their biological function, IGF-I and IGF-II exhibit significant differences in their pattern of expression *in vivo*. In particular, IGF-I is expressed in juvenile life and is synthesized almost exclusively in the liver under the control of growth hormone (as predicted by the original hypothesis). In contrast, IGF-II is expressed predominantly in the embryonic and fetal stages of mammalian development in a wide variety of different tissues. The expression of IGF-II persists into the adult only in the epithelial cells lining the surface of the brain. An additional feature of IGF-II is the existence of multiple forms of the protein differing only by a few amino acids, which are generated by both transcriptional and post-transcriptional processing events.

Unlike many of the molecules considered so far, both IGF-I and IGF-II are present in the circulation and can be readily detected in plasma. As might be

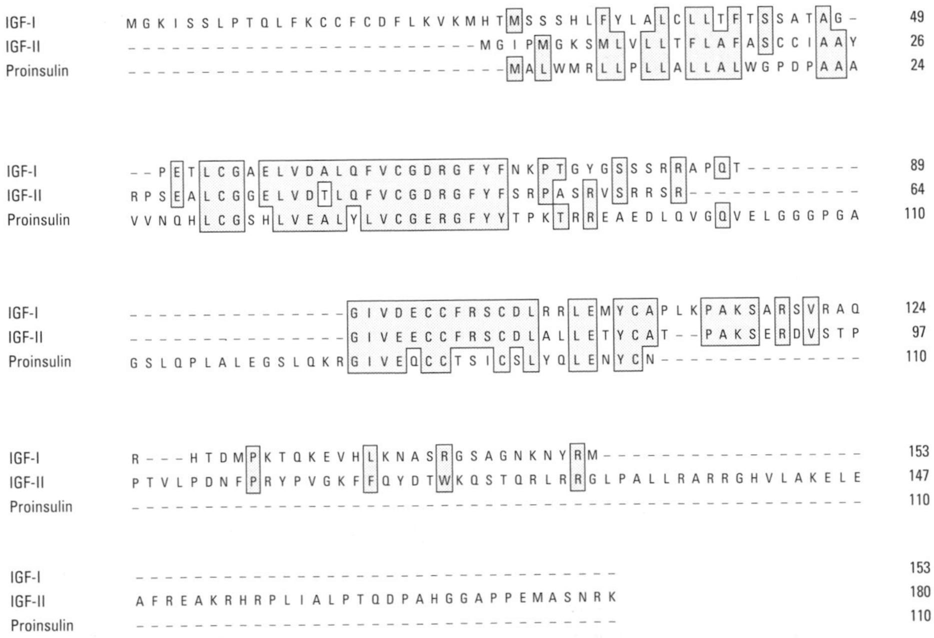

Figure 2.8 Amino acid sequence alignment of IGF-I, IGF-II and proinsulin. Regions of amino acid identity are boxed. Note the conservation of cysteine residues.

predicted from their patterns of synthesis, circulating IGF-I levels rise during juvenile life and then decline after puberty, whilst circulating IGF-II levels are highest in the fetal circulation and decline after birth. An important feature of the circulating IGFs is that they are not found free in plasma but are associated with a specific set of binding proteins. There are at least six distinct species of specific IGF-binding proteins (IGF-BPs) known and these in turn exhibit tissue-and stage-specific patterns of expression. The function of the IGF-BPs is not clear at present. *In vitro* the IGF-BPs all inhibit the biological activity of free IGFs, suggesting that part of their function may be to restrict the availability of biologically active IGFs in the circulation or locally within tissues until the IGF/IGF-BP complex is broken down by some mechanism. This does not, however, fully explain why a range of different IGF-BP forms should exist, and it may be that the differential affinity of the IGFs for individual IGF-BPs represent an additional level of control on IGF availability.

It may be thought curious that plasma has little mitogenic effect on many cell types but contains physiologically significant concentrations of the IGFs (albeit predominantly in a complexed form). Indeed in the majority of cases *in vitro* free IGFs prove to be relatively weak mitogens in their own right. The biological importance of IGF-II for growth of the whole organism can, however, be dramatically demonstrated in the living animal. Mice harbouring IGF-II genes which have been genetically inactivated exhibit significantly reduced fetal and neonatal growth rates and body mass compared to their normal wild-type counterparts (24). This convincingly demonstrates that IGF-II plays an important role in the growth of tissues in the whole organism and indicates that growth factor action in the living organism may be significantly more sophisticated than is apparent from the analysis of cell growth in culture.

3.5 The transforming growth factor (TGF) family

3.5.1 TGFβ

The preceding discussion has been concerned with molecules whose most prominent function is the ability to induce cell proliferation. There is, however, an extremely significant family of growth factors whose biological actions on cell multiplication are more subtle. In the course of the purification of TGFα from the culture supernatant of tumorigenic cells it was noted that, in the early stages of purification, fractions which contained the majority of the EGF-like bioactivity were also capable of inducing normal fibroblast cells to proliferate in semi-solid medium (soft agar) (*Figure 2.9*). This ability to proliferate without the requirement for attachment to a solid substrate has been termed 'anchorage-independent growth' and was thought to be a characteristic feature of tumorigenic (or 'transformed' cells). It came as some surprise therefore that anchorage-independent growth could be conferred on 'normal' cells by soluble polypeptides derived from the culture media of certain tumorigenic cells. This partially purified material was designated 'transforming growth factor'. However,

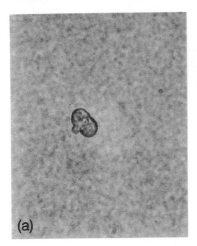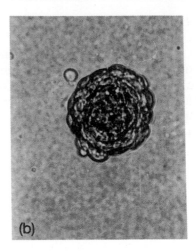

Figure 2.9.   Anchorage independent growth of rat fibroblasts in semi solid media in the presence of TGFα and TGFβ (b). (a) TGFα alone.

as further purification of this material took place it became apparent that the induction of anchorage-independent growth was the result of the dual action of two distinct polypeptides present in the impure material. One component was the EGF-like growth factor TGFα described in the preceding section, and the second ingredient was a distinct entity termed transforming growth factor beta (TGFβ). Polypeptide-induced anchorage-independent growth results from the combined action of both TGFα and TGFβ; neither agent will induce anchorage-independent growth of normal fibroblasts alone. Another way of looking at this phenomenon would be to say that TGFβ modulates the mitogenic action of TGFα such that it renders this molecule capable of inducing the multiplication of attached cells as well as inducing the growth of non-attached cells. The idea that TGFβ acts predominantly by modulating the biological effects of other growth factors (25) extends beyond the case of anchorage-independent growth.

TGFβ is a homodimeric, disulphide-bonded protein of 25 kDa made up from two 12 kDa polypeptide chains (26). Three closely related TGFβ genes exist (TGFβs 1, 2, and 3; *Figure 2.10*) creating three distinct homodimeric proteins and the hypothetical possibility of creating additional species by heterodimeric combinations of individual monomeric species. TGFβs 1–3 are widespread in their expression, TGFβ being almost ubiquitous, and in many tissues the different TGFβ family members are co-expressed. As with other growth factor families, however, it is only possible to speculate what the physiological significance of the existence of multiple, and functionally equivalent, growth factors might be. It may be important, however, that TGFβ is secreted as a proform from which the active mature protein is excised by protease action and that the three forms of TGFβ differ most significantly in the pro-region of the molecule (*Figure 2.10*).

```
TGF β2    -------MHY-CVLSTFLLLHLVPVALSLSTCSTLDMDQFMRKRIEAIR      41
TGF β3    -------MHLQRALVVLALLNLATISLSLSTCTTLDFGHIKKKRVEAIR      42
TGF β1    MPPSGLRLLPLLLPLPWLLVLTPGRPAAGLSTCKTIDMELVKRKRIEAIR     50


TGF β2    GQILSKLKLTSPPEDYPE-PDEVPPEVISIYNSTRDLLQEKASRRAAACE      90
TGF β3    GQILSKLRLTSPPEP-SV-MTHVPYQVLALYNSTRELLEEMHGEREEGCT      90
TGF β1    GQILSKLRLASPPSQGEVPPGPLPEAVLALYNSTRDRVAGESADPEPE--     98


TGF β2    RERSEQEYYAKEVYKIDMPSHLPSENAIPPTFYRPYFRIVRF----DVST    136
TGF β3    QETSESEYYAKEIHKFDMIQGLAEHNELAVCPKGITSKVFRF----NVSS    136
TGF β1    ---PEADYYAKEVTRVLMVDR---NNAIYEKTKDISHSIYMFFNTSDIRE    142


TGF β2    MEKNASNLVKAEFRVFRLQNPKARVAEQRIELYQILKSKDLTSPTQRYID    186
TGF β3    VEKNGTNLFRAEFRVLRVPNPSSKRTEQRIELFQILRPDEHI-AKQRYIG    185
TGF β1    AVPEPPLLSRAELRLQRLKSS----VEQHVELYQKYSNNSW-----RYLG    183


TGF β2    SKVVKTRAEGEWLSFDVTDAVQEWLHHKDRNLGFKISLHCPCCTFVPSNN    236
TGF β3    GKNLPTRGTAEWLSFDVTDTVREWLLRRESNLGLEISIHCPCHTFQP-NG    281
TGF β1    NRLLTPTDTPEWLSFDVTGVVRQWLNQGDGIQGFRFSAHCSC--------    263


TGF β2    YIIPNKSEELEARFAGIDGTSTYASGDQKTIKSTRKKTSGKTPHLLLMLL    286
TGF β3    DILENVHEVMEIKFKGVDNEDDHGRGDLGRLK---KQKDHHNPHLILMMI    281
TGF β1    ---DSKDNKLHVEINGISPKR---RGDLGTIH------DMNRPFLLLMAT    263
```

```
                                        ↓
TGF β2    PSYRLESQ-QSSRRKKRALDAAYCFRNVQDNCCLRPLYIDFKRDLGWKWI    335
TGF β3    PPHRLDSPGQGSQRKKRALDTNYCFRNLEENCCVRPLYIDFRQDLGWKWV    331
TGF β1    PLER--AQHLHSSRHRRALDTNYCFSSTEKNCCVRQLYIDFRKDLGWKWI    311


TGF β2    HEPKGYNANFCAGACPYLWSSDTQHTKVLSLYNTINPEASASPCCVSQDL    385
TGF β3    HEPKGYYANFCSGPCPYLRSADTTHSTVLGLYNTLNPEASASPCCVPQDL    381
TGF β1    HEPKGYHANFCLGPCPYIWSLDTQYSKVLALYNQHNPGASASPCCVPQAL    361


TGF β2    EPLTILYYIGNTPKIEQLSNMIVKSCKCS                        414
TGF β3    EPLTILYYVGRTPKVEQLSNMVVKSCKCS                        410
TGF β1    EPLPIVYYVGRKPKVEQLSNMIVRSCKCS                        390
```

Figure 2.10.   Amino acid sequence alignment of TGFβs 1, 2, and 3. The site of proteolytic cleavage of proTGFβ to release the 'mature' protein is marked. Note the high degree of conservation of amino acid sequences in the 'mature' (C-terminal) portion of the protein and conservation of cysteine residues.

A very important feature of the TGFβs is that they are secreted from cells in the form of a latent, and biologically inactive complex, formed from one molecule of TGFβ and an additional 'latency-associated protein' (*Figure 2.11*) (27). The consequence of latency is that TGFβ is biologically inactive until the latent complex is broken down. The biochemical mechanisms involved in the release from latency are uncertain but may include the involvement of either specific proteases or glycosidases. This illustrates a theme encountered in consideration of the FGFs, namely that molecules with powerful biological actions and widespread expression are held in check by mechanisms which limit their ability to encounter a responsive cell.

It has already been mentioned that TGFβs can be viewed most usefully as modulators of the action of other growth factors and this is most clearly demonstrated when considering their actions on the induction of DNA synthesis in quiescent (attached) fibroblast cells such as 3T3. If TGFβs are added to quiescent 3T3 cells they induce DNA synthesis and cell division following a significant lag period which is considerably longer than the time taken to progress through G1/S/G2/M in the presence of other growth factors such as PDGF (*Figure 2.12*). This delay is due to prolongation of G1 and it occurs because TGFβ does not directly induce the exit of the cell from the quiescent state but instead induces the synthesis of PDGF which is itself directly responsible for exit from G1 (28). The additional lag phase represents the period required for the induced expression of biologically active PDGF. On the other hand, if TGFβ is added

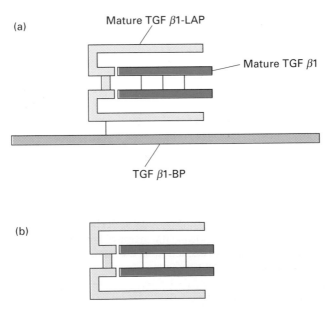

Figure 2.11.   Schematic illustration of (a) large and (b) small latent complexes of TGFβ1. LAP: latency associated protein.

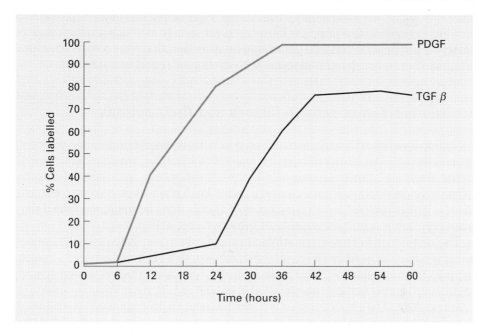

Figure 2.12. Induction of DNA synthesis in quiescent fibroblasts by TGFβ and PDGF. Note that induction of DNA synthesis by TGFβ lags behind PDGF by approximately 12 hours.

together with another growth factor, such as PDGF, to certain types of quiescent cells in monolayer it inhibits the induction of DNA synthesis, acting at some point in the cell cycle prior to the onset of DNA synthesis. In this context, therefore, TGFβ is an inhibitor of cell multiplication.

Thus it can be seen that the action of TGFβs is strictly dependent upon their context in terms of both the physiological state (and phenotype) of the responding cell and the presence of other growth factors. In some cases therefore TGFβ induces cell proliferation and in other cases cell proliferation is inhibited. The biological function of the TGFs may accordingly be complex in the whole organism and highly dependent upon the exact context in which it acts.

3.5.2 The TGFβ superfamily

TGFβs 1–3 are members of an extended 'superfamily' of structurally related molecules many of which have important biological activities. These include the activin/inhibin family composed of homo- and heterodimers of two protein chains derived from three genes α, βA, and βB the family of bone morphogenetic proteins BMPs 2–7 and mullerian inhibitory substance (MIS) (26). Additional TGFβ-like genes have been isolated from diverse species such as frogs and flies. Each of these known members of the TGFβ superfamily has biological actions distinct from both the TGFβ prototypes whilst retaining related functions within the local family. The TGFβ superfamily therefore represents a striking example of the principle introduced previously in which a large range of bioactive polypeptides may be generated by diversification from a core structural motif leading to groups of molecules with divergent regulatory functions.

4. An overview of growth factors and their actions

Attention has hitherto been focused on a relatively small number of species of growth factor. The actual number of molecules identified which exhibit growth factor activity is, however, not only large but continually increasing. Whilst an exhaustive description of every growth factor known would be beyond the scope of this chapter, at least two other classes of agents must be mentioned since they exemplify some important aspects of growth factor action.

The distinction between growth factors and other bioregulatory polypeptides becomes very blurred in the cases of small peptides such as bombesin, neuro-peptide Y, or endothelin (29). These agents, whilst having powerful and specific regulatory actions in their own right, also exhibit significant mitogenic actions on cultured cells. The importance of this will become clearer upon consideration of the biochemical action of these molecules but it draws attention again to the idea that the induction of cell multiplication is not the exclusive preserve of a discrete set of molecules but instead reflects a specific outcome which arises as the result of the interaction of many different types of molecules with many different types of cells.

The second set of growth factors is best exemplified by agents with effects on the multiplication cells of the haemopoietic system (30). Haemopoiesis involves the coordinated proliferation of multiple cell types derived from a nested set of common precursor cells under varying physiological conditions. This is achieved, it appears, by utilizing specific growth factors with specifically restricted target specificity within the diverse cell types of the haemopoietic system. The clearest example of this principle is erythropoietin which is a growth factor secreted by the kidney and acts to induce the proliferation of cells of the erythroid lineage. Other haemopoietic growth regulators exist which exhibit analogous cell-type specific restrictions in their action corresponding to different hierarchical levels of the haemopoietic system as a whole. Whilst the complexities of this system are not yet fully understood it is clear that cell-type specificity of action is the key to overall control of the whole system. The preceding sections have focused upon growth factor species which appear, predominantly from studies *in vitro*, to exhibit fairly broad-ranging actions on diverse types of cells in different tissues. The example of the haemopoietic system should be a reminder, however, that each organ in the body may be subject to equally complex and hierarchical controls involving multiple growth factor species with restricted target-cell specificities and the broad-ranging actions of many growth factors may be accordingly tightly constrained *in vivo* by control of their dissemination, availability, and delivery to responding cells.

5. Further Reading

Cross,M. and Dexter,T.M. (1991). Growth factors in development, transformation, and
tumorigenesis. *Cell* **64**, 271–80.
Sporn,M. and Roberts, A. (ed.) (1990). *Peptide growth factors and their receptors.*
Springer, Berlin.

6. References

6.1 Platelet-derived growth factor family

1. Balk,S.D. (1971) Calcium as a regulator of the proliferation of normal but not
    transformed chick fibroblasts in plasma containing medium. *Proc. Nat. Acad. Sci.
    USA* **68**, 271–5.
2. Ross,R., Raines,E., and Bowen-Pope,D. (1986). The biology of the platelet-
    derived growth factor. *Cell* **46**, 155–69.

6.2 Purification

3. Heldin,C.H., Westermark,B., and Wasteson,A. (1979). Platelet-derived growth
    factor: purification and partial characterisation. *Proc. Nat. Acad. Sci. USA* **76**,
    3722–6.

6.3 Molecular cloning

4. Johnsson,A., Heldin,C.H., Wasteson,A., Westermark,B., Deuel,T., Huang,J.,
    Seeburg,P., Gray,A., Ullrich,A., Scrace,G., Stroobant,P., and Waterfield,M.
    (1984). The *c-cis* gene encodes a precursor of the B chain of platelet-derived
    growth factor. *EMBO J.* **3**, 921–8.
5. Betsholtz,C., Johnsson,A., Heldin,C., Westermark,B., Lind,P., Urdea,M.,
    Eddy,R., Shows,T., Philptt,K., Mellor,A., Knott,T., and Scott,J. (1986). cDNA
    sequence and chromosomal localization of human platelet-derived growth factor A
    chain and its expression in tumour cells. *Nature* **320**, 695–9.

6.4 Vascular endothelial cell growth factor

6. Leung,D.W., Cachianes,G., Kuang,W.J., Goeddel,D.V., and Ferrara,N. (1989).
    Vascular endothelial growth factor is a secreted angiogenic mitogen. *Science* **246**,
    1306–9.

6.5 Epidermal growth factor family

7. Carpenter,G. and Cohen, S. (1979). Epidermal growth factor. *Ann. Rev. Biochem.*
    48, 193–216.
8. Savage,C., Inagami,T., and Cohen,S. (1972). The primary structure of epidermal
    growth factor. *J. Biol. Chem.* **247**, 7612–21.
9. Gregory,H. (1975). Isolation and structure of urogastrone and its relationship to
    epidermal growth factor. *Nature* **257**, 325–7.
10. Gray,A., Dull,T., and Ullrich,A. (1983). Nucleotide sequence of epidermal growth
    factor cDNA predicts a 128 000 molecular weight precursor. *Nature* **303**, 722–5.
11. Derynck,R. (1988). Transforming growth factor alpha. *Cell* **54**, 593–5.
12. Shoyab,M., Plowman,G.D., McDonald,V.L., Bradley,J.G., and Todaro,G.J.
    (1989). Structure and function of human amphiregulin: a member of the epidermal
    growth factor family. *Science* **243**, 1074–6.

13.   Campbell,I.,   Cooke,R.,   Baron,M.,   Harvey,T.,   and   Tapipin,M.   (1989).
      The solution structures of EGF and TGFα. *Prog. Growth Factor Res.* 1, 13–
      22.

6.6 *Fibroblast growth factor family*

14.   Burgess,W. and Maciag,T. (1989). The heparin-binding (fibroblast) growth factor
      family. *Ann. Rev. Biochem.* **58**, 575–606.
15.   Abraham,J., Whang,J., Tumulo,A., Mergia,A., Fredman,J., Gospodrawicz,D., and
      Fiddes,J. (1986). Human basic fibroblast growth factor: nucleotide sequence and
      genomic organisation. *EMBO J.* **5**, 2523–8.
16.   Jaye,M., Howk,R., Burgess,W., Ricca,G., Chiu,I., Ravera,M., O'Brien,S., Modi,
      W., Maciag,T., and Drohan,W. (1986). Human endothelial cell growth factor;
      cloning nucleotide sequence and chromosomal localisation. *Science* **233**, 541–5.
17.   Smith,R., Dickson,C., and Peters,G. (1988). Multiple RNAs expressed from the
      int-2 gene in mouse embryonal carcinoma cell lines encode a protein with homology
      to fibroblast growth factors. *EMBO J.* **7**, 691–5.
18.   Taira,M., Yoshida,T., Miyagawa,K., Sakamoto,H., Terada,M., and Sugimura, T.
      (1987). cDNA sequence of human transforming gene *hst* and identification of the
      coding sequence required for transforming activity. *Proc. Nat. Acad. Sci. USA* **84**,
      2980–4.
19.   Zhan,X., Bates,B., Hu,X., and Goldfarb,M. (1988). The human FGF-5 oncogene
      encodes a novel protein related to fibroblast growth factors. *Mol. Cell Biol.* **8**,
      3487–95.
20.   Marics,I.,   Adelaide,J.,   Raybaud,F.,   Mattei,M.G.,   Coulier,F.,   Planche,J.,   de
      Lapeyriere,O., and Birnbaum,D. (1989). Characterization of the HST-related
      FGF6 gene, a new member of the fibroblast growth factor gene family. *Oncogene*
      4, 335–40.
21.   Finch,P., Rubin,J., Miki,T., Ron,D., and Aaronson,S. (1989). Human KGF is
      FGF-related with properties of a pracrine effector of epithelial cell growth. *Science*
      245, 752–5.

6.7 *Insulin-like growth factors*

22.   Froesch,E., Schmid,C., Schwander,J., and Zapf,J. (1985). Actions of insulin-like
      growth factors. *Ann. Rev. Physiol.* **47**, 443–67.
23.   Dull,T., Gray,A., Hayflick,J., and Ullrich,A. (1984). Insulin-like growth factor gene
      organisation in relation to insulin gene family. *Nature* **310**, 777–80.
24.   DeChiara,T.M., Efstratiadis,A., and Robertson,E.J. (1990). A growth-deficiency
      phenotype in heterozygous mice carrying an insulin-like growth factor II gene dis-
      rupted by targeting. *Nature* **345**, 78–80.

6.8 *TGFβ*

25.   Sporn,M.B. and Roberts,A.B. (1988). Peptide growth factors are multifunctional.
      Nature **332**, 217–19.
26.   Massague,J., (1990). The transforming growth factor beta family. *Ann. Rev. Cell
      Biol.* **6**, 597–641.
27.   Miyazano,K., Hellman,U., Wernsetedt,C., and Heldin,C. (1988). Latent high
      molecular weight complex of transforming growth factor β1: purification from
      human platelets and structural characterisation. *J. Biol. Chem.* **263**, 6407–15.
28.   Leof,E.B.,   Proper,J.A.,   Goustin,A.S.,   Shipley,G.D.,   DiCorleto,P.E.,   and
      Moses,H.L. (1986). Induction of c-sis mRNA and activity similar to platelet-derived
      growth factor by transforming growth factor beta: a proposed model for indirect
      mitogenesis involving autocrine activity. *Proc. Nat. Acad. Sci. USA* **83**, 2453–7.

6.9 Neuropeptide growth factors

29.   Woll, P.J. and Rozengurt, E. (1989). Neuropeptides as growth regulators. *Br. Med. Bulletin* **45**, 492–505.

6.10 Haemopoietic growth factors

30.   Metcalf, D. (1991). Control of granulocytes and macrophages: molecular, cellular, and clinical aspects. *Science* **254**, 529–33.

3

Growth factor receptors and signals

1. Introduction

The induction of DNA synthesis and other biological effects of growth factors depends upon the interaction of the growth factors with specific, high-affinity receptors located in the plasma membrane of the target cells. The interaction between a growth factor and its cognate receptor on the external face of the cell triggers events inside the cell which eventually lead to the onset of DNA synthesis. The function of the growth factor is therefore activation of receptors rather than direct participation in the intracellular mechanisms of cell activation. Growth factor receptors therefore not only confer cell-type specificity on the action of growth factors but also define the biochemical identity of the subsequent intracellular events. In order to understand how growth factors work it is essential to understand the structure and function of growth factor receptors.

The molecular characterisation of a variety of different growth factor receptors reveals that they fall into defined 'families' based on their structure and biochemical actions. In many cases an individual growth factor can interact with more than one type of receptor thereby conferring additional diversity on the action of any specific growth factor. Furthermore, individual growth factor receptors may activate several different types of intracellular signalling events. The initial interaction of a single growth factor species with a responsive cell can, therefore, lead to a multiplicity of intracellular events.

The main 'families' of growth factor receptors are the tyrosine kinases, the G-protein associated seven-helical domain class, the multichain 'cytokine' family, and the serine/threonine kinases. Each family exhibits fundamental similarities in their mechanism of action which explains how different types of growth factors can elicit apparently similar cellular responses. Understanding the biochemical identity of growth factor receptor-mediated signals accordingly provides an important step in deciphering the molecular basis of the control of eukaryotic cell multiplication by intercellular signals.

2. The tyrosine kinases

Although regulation of protein function by phosphorylation is a very common general mechanism of cellular regulation, the majority of phosphorylation events involve addition or removal of phosphate groups on either serine or threonine residues. In most cell types phosphorylation of other amino acid species is rare. There is, however, one highly significant exception to this rule. Stimulation of quiescent cells by specific growth factors, such as EGF/TGFα, PDGF, or FGF leads to the rapid phosphorylation of the tyrosine residues of a specific set of proteins. Phosphorylation of tyrosine residues by tyrosine kinases can therefore be considered to be a characteristic feature of the action of certain growth factors. Molecular characterization of the cell surface receptors for growth factors such as EGF/TGFα, PDGF, and FGF reveals that within the portion of the receptor located in the cytoplasm of the cell, they all contain a polypeptide domain with tyrosine kinase activity. Binding of the growth factor to this class of receptor leads to activation of intrinsic receptor-associated tyrosine kinase activity and concomitant phosphorylation of tyrosine residues of cellular proteins. Included amongst the substrates for growth factor tyrosine kinases are the receptors themselves as well as a number of key intracellular regulatory enzymes.

2.1 The structure of tyrosine kinase receptors

The first growth factor receptor to be characterized in any detail was the cell EGF/TGFα receptor (EGF-R). This is a 130 kDa cell-surface protein which contains an extracellular domain characterized by two regions rich in cysteine, a single α-helical transmembrane domain, and a cytoplasmic domain containing the intrinsic tyrosine kinase domain and a 'tail' at the carboxy terminus (*Figure 3.1*). Addition of EGF or TGFα to cells leads to activation of the EGF-R tyrosine kinase activity and phosphorylation of a number of cellular proteins including the EGF-R itself. The majority of this 'autophosphorylation' activity of the EGF-R involves phosphorylation of tyrosine residues in the carboxy terminal domain.

   Identification and characterization of the receptors for a significant number of growth factors has revealed additional receptor species containing intrinsic ligand-activated tyrosine kinase domains (*Figure 3.2*). These include two species of PDGF receptor (PDGF-Rα and PDGF-Rβ), four species of FGF receptor (FGF-R 1–4), receptors for insulin, IGF-I/IGF-II, and receptors for a number of haemopoietic growth factors such as macrophage colony stimulating factor (M-CSF). There are, in addition, a growing number of molecules (orphan receptors) which exhibit sufficient resemblance to tyrosine-kinase-containing receptors for known growth factors to indicate that they could be receptors for as yet unidentified growth factors. An important example of this latter category is the gene *HER*-2, which is closely related to the EGF receptor.

   The EGFR can be considered, in general terms, to typify the tyrosine kinase family of receptors in terms of its modular structure—nearly all tyrosine kinase receptors described so far are composed of an extracellular ligand-binding

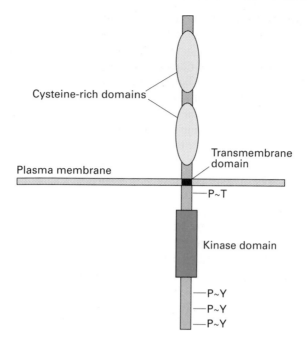

Figure 3.1. Structure of the EGF receptor: P~T shows the Thr 654 phosphorylation site and P~Y the sites of tyrosine phosphorylation.

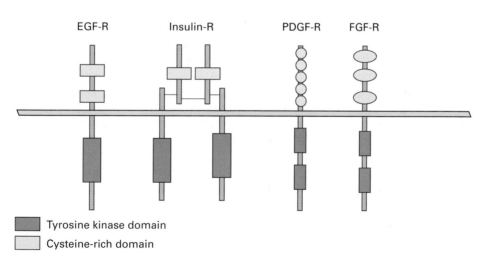

Figure 3.2. The tyrosine kinase family of growth factor receptors.

domain, a single transmembrane domain, a region containing the tyrosine kinase, and a carboxy terminal extension extending into the cytoplasm of the cell.

Within this overall picture a number of sub-categories of tyrosine kinase receptor can be identified on the basis of distinctive features of one or more of these domains. Perhaps the most significant distinction between tyrosine kinase receptors lies in the design of the tyrosine kinase domain itself. A number of tyrosine kinase receptors, including PDGF-Rα and PDGF-Rβ have 'split' kinase domains in which the region encoding the enzymatic activity is interrupted by a variable-length insertion sequence (*Figure 3.2*). The ligand-binding domain confers specificity for ligand on the receptor and is therefore of central importance in receptor function. A number of structurally related growth factors, such as EGF and TGFα, appear to interact with a unique receptor with roughly equal affinity. In other cases, as exemplified by PDGF-Rα and PDGF-Rβ, the ligand-binding domains of the respective receptors specifically bind different forms of PDGF. The design of the extracellular ligand-binding domain also varies from species to species. Whereas the EGF-R ligand-binding domain is characterized by two distinct regions rich in cysteine, in other receptors (such as PDGF-Rα and PDGF-Rβ) cysteine residues are more evenly distributed. In the FGF-R family the extracellular ligand-binding domain is related to the immunoglobulin gene superfamily in structure. Finally, whilst most tyrosine kinase receptors have a single transmembrane domain, the insulin and IGF-I/IGF-II receptors are heterodimeric molecules composed of two identical extracellular ligand-binding domains linked through disulphide bonds to a homodimeric transmembrane protein containing the tyrosine kinase domain (*Figure 3.2*).

2.2 How do tyrosine kinase receptors work?

It is clear from the presence of intrinsic ligand-activated tyrosine kinase activity in the receptors for diverse types of growth factors that tyrosine phosphorylation must play some important function in the intracellular signalling mechanism of this family of receptors. It must be presumed that activation of the kinase activity leads to tyrosine phosphorylation of certain intracellular substrates whose biological activity is thereby altered, and that the altered activity of these substrates leads, directly or indirectly, into further 'downstream' processes which ultimately result in the induction of DNA synthesis and cell division.

This overview of the action of tyrosine kinase growth factor receptors naturally leads to three key questions—what is the role of tyrosine kinase, what are the substrates for these kinases, and what activates the tyrosine kinases?

2.3 The role of the tyrosine kinases

Although it is attractive to speculate that tyrosine kinase activation is important for mitogenic signalling by growth factors it could equally be that this domain simply reflects some generic feature of this type of receptor, unrelated to its signalling function. Therefore, it needs to be established that receptor tyrosine kinase activity is intimately involved in the mitogenic action of the growth factor.

This can be directly investigated by inactivating the tyrosine kinase activity and observing the consequences on ligand-activated cellular responses.

Activation of the EGF-R by exposure of cells to EGF leads to a number of readily observable biochemical and biological events. These include relatively rapid (i.e. within a few minutes) events such as tyrosine phosphorylation of a number of cellular substrates including the EGF-R itself, activation of ion transport systems such as induction of calcium ion influx, and after a few hours, the induction of DNA synthesis followed by cell division. The binding of EGF to the EGF-R also leads to internalization of the EGF/EGF-R complex and ultimate degradation of both the ligand and the receptor. Within the tyrosine kinase domain of the EGFR are amino acid residues conserved in all protein kinases which are thought to be involved in binding the ATP substrate. By analogy with other protein kinases substitution of one such residue of the EGF-R, lysine-721, for alanine (or any other non-charged amino acid) should result in loss of the ability to bind ATP and hence inactivation of the kinase function. Transfection of such a mutated EGF-R into cells lacking a wild-type EGF-R followed by stimulation with EGF should permit determination of the requirement for tyrosine kinase activation in the induction of cellular DNA synthesis as well as other rapid and delayed responses such as substrate phosphorylation and ligand internalization. The results of such experiments are clear (*Table 3.1*)(1). Inactivation by mutation of the EGF-R tyrosine kinase destroys the ability of EGF to induce cellular DNA synthesis and substrate phosphorylation (including the EGF-R itself). Inactivation of the EGF-R does not, however, result in a loss of the receptor to bind EGF or to internalize the ligand–receptor complex. It can therefore be concluded from this and analogous experiments with other tyrosine kinase receptors, that ligand-activated tyrosine kinase activity is essential for the mitogenic action of the growth factor. This draws attention to the identity of the cellular proteins phosphorylated by ligand-activated tyrosine kinase receptors.

Table 3.1 Signalling properties of K-721A mutant EGF receptor

Ligand-activated tyrosine kinase	−
High- and low-affinity receptor subclasses	+
PKC-mediated transmodulation	+
Ligand internalization	+
Ligand degradation	+
Receptor down-regulation	−
Na^+/H^+ exchange	−
Ca^{2+} influx	−
(Inositol phosphate formation)	−
S6 phosphorylation	−
Induction of *c-fos*	−
DNA synthesis	−

2.4. Substrates for tyrosine kinases

The working hypothesis on the action of tyrosine kinase receptors described in the previous section makes certain predictions about the protein substrates involved. Firstly, such substrates should exhibit altered function following phosphorylation. Secondly, key substrates must participate in intracellular pathways leading to the induction of DNA synthesis. It is also possible to apply standard biochemical criteria to the identification of 'physiological' substrates in that they should exhibit an affinity for the enzyme which is consistent with their concentration inside the cell and their stoichiometry of phosphorylation.

The identification of substrates for tyrosine kinase receptors has proceeded along three main paths. Firstly, it is possible to label cells radioactively with ^{32}P-phosphate and, following stimulation with a growth factor, look for the appearance of protein, phosphorylated *de novo*, by gel electrophoresis (*Figure 3.3*). Such experiments readily lead to the identification of 'substrate' proteins for tyrosine kinases (2). However, in the majority of cases, it appears that proteins identified by this method fail to meet the essential criteria for a physiological substrate. The best examples are glycolytic enzymes (such as enolase) which represent the most prominent species identified by gel electrophoresis in terms of incorporation of the radiolabel. On closer inspection it emerges that these enzymes are phosphorylated to low stoichiometry and exhibit low affinity for the kinase. In addition, whilst it is conceivable that alteration of glycolytic flux is involved in the mitogenic response to growth factors, the most prominent glycolytic pathway enzymes have little impact on glycolytic flux. It seems probable in fact that the majority of proteins identified by radiolabelling and gel electrophoresis represent low affinity substrates present at high concentrations within the cell. It must be concluded, therefore, that tyrosine kinases are 'promiscuous' in terms of their substrates and that phosphorylation by a growth-factor activated tyrosine kinase cannot, *per se*, be considered proof of identification of a physiological substrate.

A more fruitful approach is to examine growth-factor activated phosphorylation of proteins thought, on the basis of independent criteria, to be involved in the mitogenic response. Four strong candidate proteins for physiological tyrosine kinase substrates have been identified by these means, each of which fit the criteria described above. The first of these is phospholipase Cγ (PLC-γ) (3), an enzyme which cleaves inositol-linked phospholipid to produce inositol derivatives and diacylglycerol. The second is an enzyme with related function, phosphoinositol 3-kinase (PI-kinase) (4), which phosphorylates the inositol phosphate sugar in the 3 position. In both cases receptor-mediated phosphorylation of the enzyme results in a distinct increase in enzyme activity *in vitro* (and presumably *in vivo*). As will become clear later from discussion of non-tyrosine kinase receptors, the identification of two tyrosine kinase substrates as mediators of inositol and diacylglycerol metabolism has considerable significance since both these molecules are involved in signal generation by other growth factors. The third protein is a threonine/serine kinase, c-raf, which was first identified as a cellular gene associated with a tumorigenic retrovirus. Phosphorylation of c-raf on tyrosine

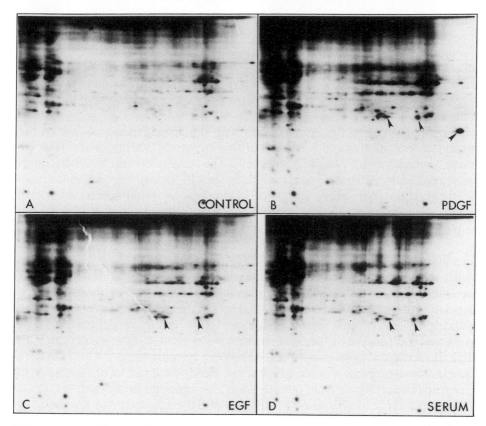

Figure 3.3.   Alkali-resistant phosphoproteins of NR-6 3T3 cells. Quiescent NR-6 3T3 cells were labelled with $^{32}P_i$ for 18 h and were either mock-treated for 60 min (a) or treated with 0.83 nM PDGF for 1 min (b), 5 min (c), or 60 min (d). Phosphoproteins were analysed on two-dimensional gels that were incubated in alkali and autoradiographed for 18 h. Arrowheads in (b), (c), and (d) show positions of phosphoproteins containing phosphotyrosine. (Reproduced from ref. 2; with permission.)

leads to activation of its intrinsic kinase activity. The physiological substrates of c-raf are currently unclear although it is reasonable to suppose that at least some of these may be involved in the further propagation of mitogenic signals into the cell. The fourth substrate for tyrosine kinase receptors is ras-GAP (5) which is a protein required for activation of the GTPase activity of the 'ras' family of membrane-associated G-proteins. The physiological function of the ras family is unclear, but it is again reasonable to suppose, by analogy with other G-protein systems, that activation of GAP by tyrosine phosphorylation leads to activation of downstream ras-mediated effector pathways.

The identification of PLCγ, PI-kinase, c-raf, and ras-GAP as high affinity receptor tyrosine kinase substrates raises the question as to how these targets are made available to the ligand-activated kinase. The answer is that each substrate appears to become physically associated with the receptor as a result of receptor autophosphorylation (6). At least in the case of the PDGF-Rα (and presumably other related receptors), the association appears to involve the 'insert region' of the split kinase domain (*Figure 3.4*). It is also notable that PLCγ and PI-kinase contain regions of homology, termed the SH2 domain (7), which may be involved in recognition of the tyrosine-phosphorylated form of the 'insert domain'.

Thus, interaction of the growth factor with the ligand-binding domain on the outside of the cell leads to activation of the enzymatic activity of the tyrosine kinase domain on the inside of the cell. The kinase domain itself becomes complexed with substrates, such as PLCγ and PI-kinase, these become activated upon phosphorylation, and these (and other) activated substrates in turn modify the behaviour of other enzyme systems within the cell, leading to the propagation of the original signal into a diversity of intracellular systems.

2.5 *Mechanisms of activating tyrosine kinase*

Whilst the outline of the basic mechanism of receptor tyrosine kinase function is becoming clear, it is important to remember that the process begins by the association of a molecule with the external face of the cell leading to change in the molecule on the internal (cytoplasmic) side. How is this brought about? The problem is complicated by the fact that, with the exception of the insulin and IGF-I/IGF-II receptors, most growth factor receptor tyrosine kinases appear to have a single transmembrane domain which is not ideally designed for transmission of a conformational change from one region of the molecule to another.

The most likely mechanism for activation of the tyrosine kinase domain is by

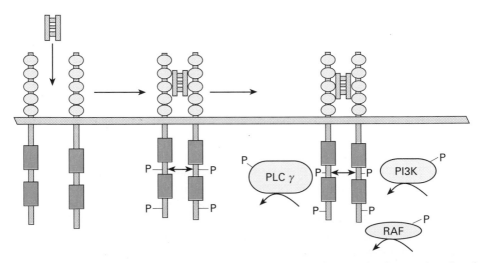

Figure 3.4.   Association of PLC, PI3 kinase, and RAF with the insert domain of PDGR-R. Autophosphorylation of receptor induces association of substrates leading to their phosphorylation and increased activity.

means of inter-receptor dimerization (8). This has been most elegantly demonstrated in the case of the PDGF-Rα and PDGF-Rβ. The two PDGF receptors exhibit different specificities for the PDGF-A and -B chains. Thus, PDGF-Rα can bind both PDGF-A and -B, whereas PDGF-Rβ can only bind PDGF-B. It has already been mentioned that PDGF is a dimeric molecule composed either of two identical A or B chains or one A and one B chain (see Section 3.1 of Chapter 2), and that one of the consequences of ligand-binding to receptor is internalization of the ligand–receptor complex. Exposure of cells which express both PDGF-Rα and PDGF-Rβ to PDGF-AA leads to internalization and down-regulation of the PDGF-Rα. This leaves only PDGF-Rβ receptors present on the cell surface. Addition of heterodimeric PDGF-AB, or homodimeric PDGF-AA, but not PDGF-BB at this point fails to generate a mitogenic signal or activate the PDGF-Rβ tyrosine kinase (9). The most reasonable explanation for this phenomenon is that activation of the PDGF-Rβ tyrosine kinase requires the formation of a receptor dimer generated by each of the two chains of the PDGF molecule, interacting with a separate receptor molecule. Since all receptors capable of binding PDGF molecules containing an A chain have been depleted, the remaining B-chain specific PDGF-Rβ molecules are unable to respond to PDGF-AB (or AA) since they are unable to form a dimer. Dimers can still be formed, however, with PDGF-BB (*Figure 3.5*). In an analogous fashion, PDGF heterodimers in which one chain has been mutated such that it cannot bind the receptor can block the effects of native PDGF containing two ligand-binding chains. This occurs because PDGF receptor occupancy by heterodimeric mutant PDGF blocks the ability of receptors to associate into dimers.

The philosophy of this experiment can be carried one stage further. The

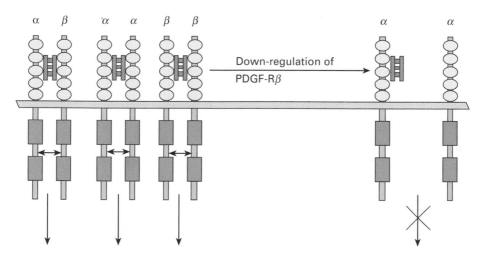

Figure 3.5.   Dimerization of PDGF receptors is required for function. The β receptor binds only B chains whereas the α receptor binds both A and B. Down-regulation of α receptors by exposure to AA homodimer means that PDGF-AB is no longer active since both α and β receptors are required to form a dimer.

preceding experiment leads to the conclusion that PDGF-mediated dimerization of ligand-binding domains on the outside of the cell brings the tyrosine kinase domains of each receptor into close proximity. Transfection of mutant PDGF receptors lacking a tyrosine kinase domain into cells which already express functional PDGF receptors blocks the tyrosine kinase activity (and mitogenic signalling) of the resident receptors. This can be explained by the formation of dimers between active and inactive receptors which are rendered inactive if one, or both, partners in the complex lack the tyrosine kinase domain (*Figure 3.6*). This shows that activation of tyrosine kinase activity requires intermolecular association of the cytoplasmic domains of the receptors, and if one of the partners in the complex is non-functional then the tyrosine kinase cannot be activated. This finding accordingly provides an explanation for the phenomenon of autophosphorylation of growth factor tyrosine kinase receptors—that is, it represents intermolecular phosphorylation resulting from tyrosine kinase activation following receptor dimerization.

Growth factor tyrosine kinase receptors lacking functional tyrosine kinase domains, or PDGF molecules containing a mutated chain, accordingly exhibit 'dominant negative' activity in that they block the signalling capacity of resident wild-type receptors. Such dominant negative mutants represent a potentially powerful means for analysing the biological function of PDGF *in vivo*.

3. G-protein-linked receptors

In the previous chapter it was noted that a number of molecules initially identified as neuropeptides or vasoactive agents could be demonstrated to exhibit growth-factor-like mitogenic effects on cultured cells. Examples of such molecules

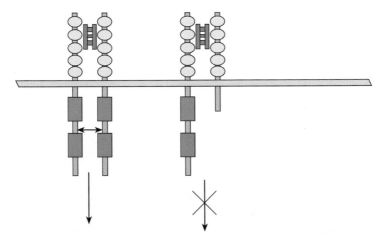

Figure 3.6 Dominant negative mutants of the PDGF-R. Deletion of the tyrosine kinase domain results in receptors which are able to undergo ligand-induced dimerization but are unable to signal.

include bombesin, bradykinin, and the endothelins. Molecular cloning of the receptors for these and other growth factors reveals that they belong to a family of receptors distinct from the tyrosine kinases. This second family of receptors resemble the light-activated receptor rhodopsin and the α-adrenergic receptor in structure, being composed of multiple (usually seven) α-helical membrane-spanning domains linked by polypeptide loops (*Figure 3.7*). The structural analogy with the well characterized examples of rhodopsin and β-adrenergic receptor suggest that they act by means of an analogous signal-transduction system, namely the use of 'G-proteins' to couple receptor activation to 'downstream events'. In the case of the endothelin, bombesin, and bradykinin and related mitogenic receptors the coupled 'downstream event' is the hydroly-sis of membrane phosphatidylinositol lipids by phospholipase C, liberating inositol polyphosphates and diacylglycerol (DAG). Thus, although the design of the G-protein linked receptors is quite distinct from the tyrosine kinases, both families appear to activate, directly or indirectly, cellular processes mediated by DAG and inositol polyphosphates. The major biochemical function of DAG is the activation of a key signalling enzyme, protein kinase C.

4. Protein kinase C

Protein kinase C (PKC) is a multigene family of closely related membrane-associated, calcium-activated, anionic phospholipid-dependent serine/threonine protein kinases. In the presence of DAG the calcium ion concentrations required for enzyme activation drop to those found within the cell, leading to induction of

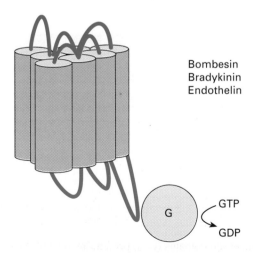

Figure 3.7   The structure of G-protein-linked receptors. Receptors are composed of multiple membrane-spanning domains linked by polypeptide loops. G-proteins are associ-ated with the receptor on the cytoplasmic face of the receptor. Binding of ligand to the external domain induces activation of the GTPase activity of the G-protein.

PKC activity and phosphorylation of cellular protein substrates. Anionic phospholipids appear to be required for DAG to bind to PKC. The structure of the PKC family (*Figure 3.8*) reveals a number of conserved domains. The C terminus contains a catalytic domain with sequence similarity to other protein kinases. The catalytic domain has constitutive protein kinase activity when the N-terminal region of the protein is deleted, suggesting that the N-terminus (or regulatory) domain has some inhibitory or controlling function in the context of the complete molecule. In addition, the regulatory domain contains sites for interaction with DAG, anionic phospholipid, and calcium ions. Within the N-terminal domain of all PKCs is a pseudo-substrate region, C1, whose deletion leads to kinase activation and which is presumably directly responsible for enzyme inhibition, and a domain found in some, but not all PKCs, C2, which confers calcium-dependence on enzyme activity. The presence or absence of the C2 domain in PKC family members indicates that at least one difference between different PKCs lies in differential sensitivity to calcium ion concentration for activity and it is therefore noteworthy that PKC family members lacking C2 domains predominate in tissues such as skeletal muscle in which changes in intracellular calcium concentration are linked to differentiated function.

There is very good experimental evidence that activation of PKC mediates mitogenic responses. The fact that the action of a number of growth factors (especially those using G-protein-linked receptors but also tyrosine kinase dependent growth factors such as PDGF) includes activation of PKC points to the possibility that PKC activity has some important function in the mitogenic response. The major evidence, however, arises from the discovery that a class of lipids, phorbol esters, (or tumour promoters by virtue of their ability to

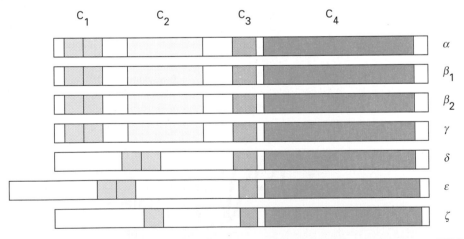

Figure 3.8   The conserved domains (C_{1-4}) present in most of the currently known PKC gene products are shown. All these PKCs show essentially colinear kinase domains spanning C_3 and C_4. By contrast the regulatory domains are more varied with no C_2 domain present in δ, ε, or ζ. All these polypeptides do nevertheless contain at least one of the cysteine-rich regions present in the first conserved domain (C_1).

accelerate the production of carcinogen-induced tumours in experimental animals) act as non-metabolizable analogues of DAG and cause prolonged activation of PKC when applied to quiescent cells. Phorbol esters are powerful mitogenic agents in their own right and the concentration of any individual phorbol ester required to induce DNA synthesis and cell division roughly corresponds to its affinity for PKC. Phorbol esters can therefore be considered to be 'pharmacological mitogens' whose sole action appears to be activation of PKC. Furthermore, it is possible to demonstrate that membrane-permeable analogues of DAG are also mitogenic for quiescent cells if added in sufficient amounts to prevent depletion by cellular DAG-metabolizing enzymes. Finally, elevation of PKC activity by overexpression of PKC in cells under the control of a constitutive promoter can lead to prolonged stimulation of cell multiplication.

PKC activation can therefore be confidently identified as a significant event for the induction of DNA synthesis in quiescent cells. By analogy with tyrosine kinases, it can be argued that activation of PKC by means of activation of PLCs, leads to phosphorylation of essential cellular substrates for PKC, whose altered activity leads, directly or indirectly, to cellular events controlling the exit from G1 and entry into DNA synthesis.

Progress in the identification of physiological PKC substrates (employing analogous criteria to tyrosine kinase substrates) has, however, been slow. The majority of protein substrates of PKC characterized so far are membrane receptors for other ligands, including the EGF-R. The effect of EGF-R phosphorylation by PKC is to induce a loss of affinity in a subclass of 'high affinity' EGF-Rs found in many cell types. The role of these high-affinity receptors in the mitogenic action of EGF is, however, unclear. The modulation of receptor affinity for heterologous ligands by means of phosphorylation by PKC, appears to be a common manifestation of PKC action which indicates that its actions are pleiotropic and extend from direct activation of mitogenic signalling pathways into modulation of a wide variety of receptor-mediated cellular functions.

It will be noted that activation of protein phosphorylation is an important component of the intracellular action of growth factors. It is very likely, from analogy with control of enzyme activity by phosphorylation characterized in other enzyme systems, that the phosphorylation state of key substrates is controlled both by the action of tyrosine and serine/threonine kinases and by the activity of specific phosphatases. In particular it is clear there are specific tyrosine phosphatases whose activity counters the action of tyrosine kinases. It is not, however, clear how the action of these enzymes is regulated or what role they might play in attenuation or modulation of growth factor mediated mitogenic signals.

5. Inositol polyphosphates

The second 'arm' of the action of G-protein-linked growth factor receptors is the release of inositol polyphosphates from the hydrolysis of phosphatidylinositol. The major target for G-protein-linked PLCs is PIP2 (phosphatidylinositol 4,5 bisphosphate). Hydrolysis of PIP2 releases DAG and inositol 1,4,5 trisphosphate (IP3). IP3 thus released can enter a number of metabolic pathways including additional phosphorylation to IP4, IP5, and IP6 or dephosphorylation through a number of biochemical routes to inositol from which inositol phospholipid may be resynthesized (*Figure 3.9*). The rapid metabolism of IP3 into higher or lower phosphorylated forms means that receptor-mediated rises in intracellular IP3 levels are typically transient in nature and subside to resting levels within a few minutes of stimulation by ligand. In addition, increasing evidence suggests that activation of PKC induces activation of PLCs which hydrolyse other species of membrane phospholipid, principally phosphatidylcholine, which results in sustained release of DAG without accompanying IP3.

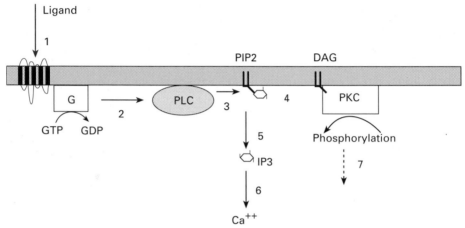

Figure 3.9.   The two branches of the phosphatidylinositol pathway. The interaction of a growth factor with its receptor (1) leads to binding and activation of a G protein which dissociates from the receptor and activates phospholipase C (2). Activation of phospholipase C leads to hydrolysis of phosphatidylinositol biphosphate (PIP2) (3). Hydrolysis of PIP2 leads to release of diacylglycerol (4) and IP3 (5). The resulting elevation of intracellular IP3 leads to release of Ca^{++} from intracellular stores (6). Diacylglycerol in the presence of Ca^{++} and phospholipid activates protein kinase C leading to phosphorylation of cellular substrates (7).

The functions (if any) of many inositol polyphosphates are currently obscure. IP3, however, has a clearly defined function which is the release of calcium from intracellular stores through association with a specific 'receptor' located in the endoplasmic reticulum. A burst of increased intracellular free calcium is therefore characteristic of the action of growth factors such as bombesin or endothelin whose major activities arise through hydrolysis of PIP2. The function of IP3 release and the associated rise in intracellular calcium in terms of the mitogenic response is, however, less clear. It is conceivable, for example, that a transient rise in free 'calcium' might cause activation of calcium-dependent protein kinases or more global changes in the intracellular concentrations of other ionic species. However, the artificial induction of a rise in intracellular free calcium by means of ionophores rarely results in longer term effects on the entry into DNA synthesis or cell division, although calcium ionophores can, in certain specific circumstances, enhance the mitogenic effects of phorbol esters. Thus it can be concluded that IP3 release (and the concomitant rise in free calcium), whilst a diagnostic aspect of G-protein-linked growth factor receptors, is not directly linked to the mechanism of exit from G1 and the entry into DNA synthesis but is instead involved in activating other classes of cellular response.

6. The cytokine superfamily of receptors

Molecular cloning and characterization of receptors for a number of haemopoietic growth factors including erythropoietin, granulocyte colony stimulating factor (G-CSF), granulocyte macrophage colony stimulating factor (GM-CSF), and interleukins 2,3,4,5,6,7 has revealed the existence of another 'superfamily' of growth factor receptors sharing common structural features (*Figure 3.10*). The diagnostic feature of the 'cytokine' receptors is the presence, in the extracellular region of the receptor, of a domain containing multiple cysteine residues and a conserved amino acid motif WSXWS (Trp–Ser–X–Trp–Ser). Mutagenesis studies of this domain reveal that its main function is recognition and binding of the ligand. Many of these receptors also contain additional extracellular domains related in sequence to the 'type III repeats' of the cell-adhesion protein fibronectin. The cytoplasmic regions of the cytokine family are typically very short and do not contain any sequences suggestive of, for example, protein kinase activity. An

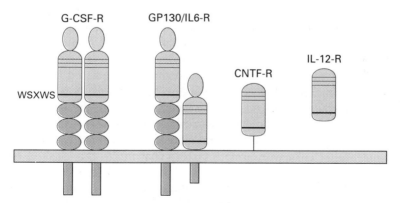

Figure 3.10.   Some members of the 'cytokine' family of receptors. These are characterized by the presence of a ligand-binding domain containing multiple cysteine residues (bars) and the characteristic WSXWS amino acid motif. In some cases, such as G-CSF-R, the active form of the receptor is a homodimer, in others such as IL-6-R, binding the ligand induces association with a second 'transducing chain' gp130. All cytokine family receptors characterized to date have short cytoplasmic domains. In the case of CNTF-R the ligand-binding module is linked to the membrane by a glyco-inositol lipid anchor. In the case of IL-12 the ligand-binding domain is a secreted molecule. In these cases association with a second, signalling, chain must be required to generate an intracellular signal.

additional typical feature of this family is the presence of 'soluble receptors' often generated by alternative RNA splicing mechanisms. These soluble receptors lack the transmembrane and cytoplasmic domains and appear as though they were secreted forms of the ligand-binding domain. In two cases, ciliary neuro-trophic factor (CNTF) and interleukin-12, the ligand-binding domain is either tethered to the membrane by a glycoplipid anchor (CNTF) or is recovered complexed with the ligand (IL-12) (*Figure 3.10*).

These findings immediately raise the issue of how these agents generate a mitogenic signal; their receptors have either very short or completely absent cytoplasmic domains. It appears that they generate a mitogenic signal by interaction of the receptor–ligand complex with an additional transmembrane molecule whose primary function is signal transduction. The clearest example of this is interleukin-6 (IL-6) where signal transduction requires an additional 'signal converting' molecule gp130 (10), which is itself a member of the cytokine receptor family although apparently unable to bind IL-6 directly. Thus IL-6 bound to the IL-6 receptor is mitogenically inactive in the absence of gp130 and IL-6 mediated signal transduction is initiated by the formation of a physical complex between the IL-6/IL-6-R complex and gp130 (*Figure 3.1*). This action, by means of subunit dimerization, is reminiscent of the tyrosine kinase receptors, and it can be presumed that, in the cases where currently characterized receptors lack transmembrane or cytoplasmic domains, an additional transmembrane signal-converting molecule remains to be identified. Moreover, an intriguing feature of certain members of the cytokine family is that they appear to utilize a common signal-converting chain. Thus the GM-CSF, IL-3, and IL-5 receptors all interact with a common transmembrane 'β-subunit' molecule AIC2B to form a high-affinity signalling complex in the presence of ligand. This suggests that the ligand-binding receptor confers specificity in terms of identification of the ligand but that each of these growth factors activates similar, if not identical, intra-cellular signal transduction pathways.

The molecular mechanism of signal transduction mediated by the cytokine family is not, at present, wholly clear. It seems highly likely, however, that these receptors act through tyrosine kinases which are either tethered to the internal face of the cell membrane or cytoplasmic in location. The evidence for this is indirect in that rapid tyrosine phosphorylation of cellular substrates is often found after stimulation of cells with growth factors that utilize this class of receptor, and in the case of the receptor for interleukin-2 a cytoplasmic tyrosine kinase, lck, has been found associated with the ligand–receptor complex after stimulation.

7. The serine/threonine kinase receptors

Very little is known about the molecular nature of the receptors for the TGFβ superfamily of growth factors. However, molecular cloning of a high-affinity receptor for activin has uncovered a potentially novel class of receptor which is characterized by a cytoplasmic domain containing a region encoding serine/threonine kinase activity (11). The activin receptor is related in sequence to phylogenetically conserved transmembrane serine/threonine kinases identified in various invertebrate species. Although nothing is currently known about the substrates for these receptor kinases it seems very likely, based on understanding of other growth factor receptors, that other members of the TGFβ superfamily will prove to act through a related set of receptors.

8. Divergent and convergent pathways

So far the mechanisms of activation of different types of growth factor receptors by their cognate ligands leading to a multiplicity of cellular responses of both long term and short term nature have been discussed. It seems highly significant that both the tyrosine kinase and G-protein type receptors (and most probably the cytokine and serine/threonine kinase families) output into cellular systems which involve protein phosphorylation as their central mode of action. Within this generalization it is also apparent that activation of PKC, by means of PLC-mediated release of DAG, is a common theme in many of the cases we have considered so far. Thus different types of growth factor, acting through different types of specific receptors, can be considered to exhibit convergent functions—their diverse actions lead to a small set of cellular systems involved in the mitogenic response.

By the same argument every individual growth factor receptor displays divergent functions, principally (through the ligand-binding domain) in 'programming' the specificity of a cell's ability to respond to particular growth factors. Growth factor receptors also display divergent actions at the level of the intracellular response either, in the case of tyrosine kinase receptors, through substrate specificity or, in the case of G-protein-linked receptors, by linking the activation of a mitogenic pathway with other metabolic pathways with distinct functions. Thus it begins to become apparent that growth factors, by means of these divergent pathways, can elicit a wide array of biological responses aside from the induction of DNA synthesis.

9. Further reading

9.1 Structure/function of tyrosine kinases: reviews

Yarden,Y., and Ullrich,A. (1989). Growth factor receptor tyrosine kinases. *Ann. Rev. Biochem.* **57**, 443.
Ullrich,A., and Schlessinger,J. (1990). Signal transduction by receptor with tyrosine kinase activity. *Cell* **61**, 203.

9.2 Phosphatidyl inositol and protein kinase C reviews

Berridge,M.J. (1987). Inositol triphosphate and diacylglycerol: two interacting second messengers. *Ann. Rev. Biochem.* **56**, 159–94.

Coussens, L., Parker, P.J., Rhee, L., Yang-Feng, T.L., Chen, E., Waterfield, M., Francke, U., and Ullrich, A. (1986). Multiple, distinct forms of bovine and human protein kinase C suggest diversity in cellular signaling pathways. *Science* **233**, 859–66.

Nishizuka, Y. (1988). The molecular heterogeneity of protein kinase C and its implication for cellular regulation. *Nature* **334**, 661–5.

9.3 G-protein receptors: review

O'Dowd, B., Lefkowitz, R., and Caron, M. (1989). Structure of the adrenergic and related receptors. *Ann. Rev. Neurosci.* **12**, 67–85.

9.4 Cytokine superfamily: review

Bazan, J. (1990). Structural design and evolution of a cytokine receptor superfamily. *Proc. Nat. Acad. Sci. USA* **87**, 6934–8.

9.5 Protein phosphatases

Tonks, N.K. (1990). Protein phosphatases: key players in the regulation of cell function. *Curr. Opin. Cell Biol.* **2**, 1114–24.

10. References

10.1 Receptor mutants

1. Honneger, A., Dull, T., Felder, S., and Schlessinger, J. (1987). Point mutation at the ATP binding site of EGF receptor abolishes protein tyrosine kinase activity and alters cellular routing. *Cell* **51**, 199–209.

10.2 Tyrosine kinase substrates

2. Cooper, J.A., Bowen-Pope, D.F., Raines, E., Ross, R., and Hunter, T. (1982). Similar effects of platelet-derived growth factor and epidermal growth factor on the phosphorylation of tyrosine in cellular proteins. *Cell* **31**, 263–73.
3. Wahl, M., Nishibe, S., Suh, P., Rhee, S., and Carpenter, G. (1989). Epidermal growth factor stimulates tyrosine phosphorylation of phospholipase Cα independently of receptor internalisation and intracellular calcium. *Proc. Nat. Acad. Sci. USA* **86**, 1568–72.
4. Coughlin, S., Escobedo, J., and Williams, L. (1989). Role of phosphatidylinositol kinase in PDGF receptor signal transduction. *Science* **243**, 1191–5.
5. Ellis, C., Moran, M., McCormick, F., and Pawson, T. (1990). Phosphorylation of GAP and GAP-associated proteins by transforming and mitogenic tyrosine kinases. *Nature* **343**, 377–81.
6. Kazlaukas, A. and Cooper, J. (1989). Autophosphorylation of the PDGF receptor in the kinase insert region regulates interaction with cellular proteins. *Cell* **58**, 1121–33.
7. Koch, C.A., Anderson, D., Moran, M.F., Ellis, C., and Pawson, T. (1991). SH2 and SH3 domains: elements that control interactions of cytoplasmic signaling proteins. *Science* **252**, 668–74.

10.3 PDGF receptors

8. Heldin, C. and Westermark, B. (1989). Platelet-derived growth factor: three isoforms and two receptor types. *Trends Genetics* **5**, 108–11.

9. Hammacher,A., Mellstrom,K., Heldin,C.H., and Westermark,B. (1989). Isoform specific induction of actin reorganisation by platelet-derived growth factor suggests that the functionally active receptor is a dimer. *EMBO J.* **8**, 2489–95.

10.4 Gp130 and interleukin-6

10. Hibi,M., Murakami,M., Saito,M., Hirano,T., Taga,T., and Kishimoto,T. (1990). Molecular cloning and expression of an IL-6 signal transducer gp130. *Cell* **63**, 1149–57.

10.5 Serine/threonine kinase receptors

11. Mathews,L.S. and Vale,W.W. (1991). Expression cloning of an activin receptor, a predicted transmembrane serine kinase. *Cell* **65**, 973–82.

4

Growth factors and the nuclear response

1. Introduction

Growth factors exert their biological effects on cells by interaction with specific cell surface receptors leading to the activation of a number of possible signal transduction pathways. The endpoint of growth factor action resides, however, in the regulation of gene expression. Most of the major biological consequences of growth factor action, including the induction of DNA synthesis, arise directly or indirectly as a consequence of short or long term changes in gene expression.

Activation of gene expression by growth factor-mediated signals has a number of characteristic features. A specific subset of genes are activated or repressed and the exact set (or combination) of genes responsive to any growth factor is highly influenced by the identity of the responding cell. The effects on gene expression are not instantaneous but occur at varying times after the initial interaction of the growth factor with its receptor in a manner which depends upon the identity of the responsive gene. Furthermore, a large proportion of the genes whose expression is affected by growth factor action are unrelated to the processes leading to the induction of DNA synthesis but reflect more widespread manifestations of growth factor action on cell behaviour, phenotype, and differentiation. A subset of growth-factor regulated genes encode transcriptional regulatory factors, and some of these factors have been strongly implicated in the mechanism of exit from G1 and induction of DNA synthesis.

2. The requirement for gene expression

It is not necessarily self-evident that the action of growth factors should involve, as an essential aspect, regulation of gene expression. Consideration of the experiments leading to the postulation of the restriction point (Chapter 1) indicate, however, that growth factors are not required to be present throughout the life-time of a cell in order to exert their effect but, specifically in terms of the induction of DNA synthesis, growth factors are required for a defined period in

57

G1 after which further progress through the cell cycle becomes independent of the exogenous signal. These experiments could be interpreted to suggest that the growth factor induces some type of stable change in the cell after an initial period of activity.

It can be demonstrated that this process of induction requires both protein synthesis and gene transcription by the following experiment (*Figure 4.1*) (1). Quiescent cells are 'primed' by exposure to PDGF for a period of time sufficient to commit the cell to enter DNA synthesis and the PDGF is then removed. Before DNA synthesis commences the cells are enucleated and fused with 'virgin' nuclei from a parallel set of cells which have never been exposed to PDGF. It is observed that the cytoplasm of cells exposed to PDGF are able to induce DNA synthesis in the nuclei of cells that have never been directly exposed to PDGF. In other words, exposure to PDGF has generated a 'cytoplasmic state' which is capable of inducing DNA synthesis upon a heterologous nucleus and for which the continued presence of PDGF is not required. If, however, the cells exposed to PDGF are treated with inhibitors of either RNA transcription or protein synthesis during the period of 'priming', the ability to induce DNA synthesis in the virgin nucleus is abrogated. It may therefore be concluded that the process of 'priming' during exposure to PDGF, which is manifest by the ability to induce entry into S phase in a nucleus from an unprimed cell, requires gene transcription and protein synthesis. Since the induction of DNA synthesis in the donor nucleus is insensitive to inhibition of transcription or translation it may also be concluded that the effects of these inhibitors involve genes which are transcribed or translated as a result of exposure to PDGF.

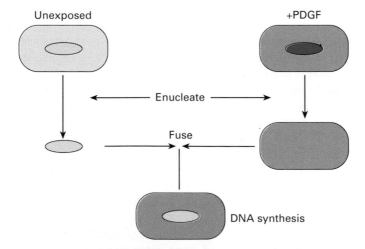

Figure 4.1 Induction of DNA synthesis in virgin nuclei by fusion with cytoplasts previously exposed to PDGF (after Smith and Stiles, ref 1). Quiescent fibroblasts are exposed to PDGF for 2 hours and then enucleated. The resulting cytoplasts are then fused with nuclei from cells that have not been exposed to PDGF and induction of DNA synthesis in the hybrid cells measured by incorporation of radiolabelled thymidine.

3. Growth-factor regulated genes

The demonstration that *de novo* gene expression is required for the mitogenic effects of growth factors naturally focuses attention on the identity of the relevant genes. In a similar fashion to the identification of kinase substrates, there are broadly two approaches to this issue. In the first case it is possible to isolate genes, by a number of technical strategies, simply on the basis of their differential expression in cells exposed to growth factors compared to quiescent cells (rather than by some feature of their DNA sequence) (2). The alternative strategy is to examine the expression of known genes, in quiescent cells exposed to growth factors, whose identity suggests that they may have some role to play in the biological response to growth factors (3).

Both the above strategies have borne fruit and there are now a considerable number of genes which are known to be transcriptionally activated in quiescent cells after exposure to a growth factor signal. Inspection of a list of only a few of these genes (*Table 4.1*) reveals some important aspects. Firstly, whilst some genes can be clearly implicated in the processes of DNA synthesis (such as nucleotide metabolism enzymes or histones) a considerable proportion appears to have biological functions unrelated to the processes of mitogenesis. For example, represented amongst the set of growth-factor-inducible genes are several secreted extracellular proteases and extracellular matrix components. This again illustrates the point that growth factors may have many important biological effects on cell function aside from the induction of DNA synthesis. Secondly, it is not necessarily trivial that most genes do not exhibit changes in expression as a result of growth factor stimulation. Thirdly, a significant fraction

Table 4.1   Some examples of growth-factor inducible genes

Early genes	Function
c-fos	
Krox-20	
Krox-24	Transcription factors
Fra-1	
c-myc	
Intermediate genes	
Collagenase	Metalloprotease
JE	Cytokine
Cathepsin L	Protease
Osteopontin	Extracellular matrix protein
Fibronectin	Extracellular matrix protein
TIMP	Protease inhibitor
Late genes	
Dihydrofolate reductase	Nucleotide metabolism
Histone H4	Chromatin structure
Thymidine kinase	Nucleotide metabolism

of growth-factor-responsive genes, in particular those with clear non-mitogenic functions, are only induced in response to growth factor stimulation in a selected set of target cells. Thus the extracellular matrix protein, fibronectin, may be induced in some cell types but not in others. Fourthly, many growth-factor inducible genes can be induced in responsive cells by a variety of different growth factors as well as pharmacological mitogens such as phorbol esters. The regulation of gene expression therefore often represents a generic response to growth factors which act through different classes of receptor and signal transduction systems.

The conclusion is drawn therefore, that the induction of gene expression by growth factors involves a select group of genes, representing a wide variety of biochemical functions whose expression is often subject to some form of cell-type specific programming. It is also evident that the level of gene expression represents a cellular response in which the actions of different growth factors converge.

4. Timing of the genetic response

Examination of the longitudinal response of a number of growth-factor inducible genes in a quiescent cell exposed to a growth factor shows that growth-factor inducible genes are not induced in parallel but rather that each gene displays a characteristic 'timing' of induction following exposure of the cell to the growth factor. These can be roughly divided into 'early genes' which are rapidly induced (within minutes), 'intermediate' genes (induced within a few hours), and 'late' genes (induced after a few hours). Thus whereas activation of signal transduction tems such as tyrosine phosphorylation, or activation of PLCs occurs as soon as the growth factor associates with the receptor, the genetic response may be delayed by many hours. This suggests that some time-dependent process is involved in the mechanism of activation, particularly in the case of intermediate and late genes. A possible explanation for this effect is that the induction of intermediate genes is, in some way, dependent upon the induction of prior early genes. There is, in other words, a cascade of gene activation in which some genes are expressed rapidly in response to receptor-mediated signalling and that at least some of the early class of genes are involved in the transcriptional activation of the intermediate class, some of which in turn are involved in the late class of gene activation. It will be apparent immediately that this provides a potential mechanism by which the programmed longitudinal activation of cellular responses may become uncoupled from the initial stimulus (*Figure 4.2*). It follows from this model that at least some of the early class of genes should include transcriptional activators of the intermediate class, and the 'timing' of genetic responses reflects the period required for functional transcriptional activators to be produced and act on their intermediate targets. In the context of induction of DNA synthesis, and the molecular mechanism of exit from quiescence, it might also be concluded that amongst the class of intermediate genes

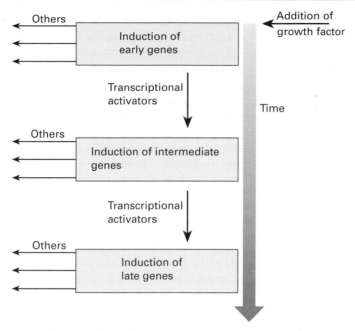

Figure 4.2   Cascade regulation of gene expression. Some rapidly induced genes induce the expression of transcription factors which in turn induce the expression of further genes.

are those involved in activating the expression of 'late' genes involved in the actual machinery of DNA synthesis.

5. The regulation of intermediate gene expression

The model described can be investigated in several ways. Amongst the most fruitful approaches has been the analysis of *cis*-acting transcriptional regulatory elements required for intermediate gene expression. An example intermediate gene is the secreted metalloproteinase collagenase whose expression can be induced in quiescent fibroblasts by a variety of growth factors as well as (significantly) phorbol esters such as TPA. Analysis of sequences present in the collagenase promoter required for transcriptional activation by TPA by un-covered a short region (termed the TRE or TPA response element) (4) which could confer induction, with intermediate timing characteristics, by TPA (or growth factors such as FGF) on a heterologous gene. Furthermore many (but not all) growth-factor inducible genes with intermediate timing characteristics contained identical or related sequences within their genetic regulatory elements. This suggests that activation of many intermediate genes involves a molecular species which interacts, in a sequence-specific manner, with the TRE (or related sequences) whose function is to induce transcription of the target gene.

A variety of different technical approaches led to the identification and characterization of the transcriptional activator responsible for TRE-dependent growth-factor-mediated gene expression. This proved to be a complex of two proteins, c-fos and c-jun, which act by binding directly to the TRE and closely related sequences (5, 6). c-fos and c-jun interact to form a transcription complex by means of a protein structural motif termed a 'leucine zipper'. This is created by an α-helical domain containing a regular spacing of leucine residues on one face of the helix (*Figure 4.3*). The leucine residues, in association with specific surrounding amino acids, permit the two molecules to dimerize creating the sequence-specific DNA-binding domain (7). A variety of evidence demonstrates that sequence-specific binding of the fos/jun dimer initiates gene transcription, and perhaps the most compelling evidence is that fos/jun-mediated TRE-dependent transcription is abolished upon mutation of the leucine zipper domain of either partner.

There are two significant features of *c-fos* and *c-jun* expression. The first is that both genes can be shown, in the presence of the cognate partner, to induce (continued) cell multiplication in appropriate target cells. Both these genes are therefore intimately involved in the molecular mechanisms of cell proliferation. Secondly, *c-fos* is a gene whose expression is induced (with 'early' kinetics) in response to growth factor stimulation of a wide variety of cell types. Thus TRE-dependent activation of intermediate genes, such as collagenase, arises (at least in part) from the prior induction of c-fos which associates with c-jun to activate gene expression. The biological consequences of constitutive co-expression of *c-fos* and *c-jun* indicate that, amongst the intermediate class of growth factor-

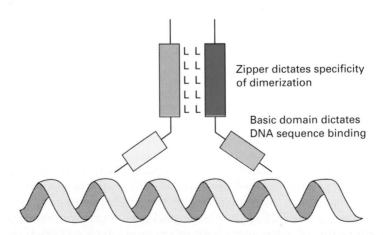

Figure 4.3.   The leucine zipper motif found in a number of growth-factor inducible transcription factors. The presence of an α-helical domain with leucine residues spaced down one face permits the formation of homo- and heterodimers. The creation of dimers brings the basic domains of each partner into alignment creating a sequence-specific DNA-binding domain.

inducible genes, are fos/jun-dependent genes involved in further propagation of the mitogenic response.

6. What activates the activator and other problems

The example of *c-fos* and *c-jun* provides a satisfying illustration of the idea of a timed cascade of gene activation underlying the mitogenic response to growth factors. At the same time, however, it poses a series of problems, none of which, at the present time, can be fully answered.

The first problem lies in the identity of the 'downstream' targets of fos/jun action. It is axiomatic from the discussion so far that genes activated by fos and jun must include those which act directly or indirectly, to induce entry into DNA synthesis, perhaps by inducing transcriptional activation of key parts of the synthetic machinery. The precise identity of such hypothetical genes is not, however, clear.

The second issue arises from consideration of the mechanism of 'uncoupling' gene expression from the initiating stimulus. The biological evidence all points to the existence of a period when growth factor/receptor signalling is required, leading into a phase in which cell-cycle events occur without the requirement for a continued signal. An attractive explanation for this phenomenon is the idea of 'autoactivation', namely that amongst the targets for a growth-factor inducible transcriptional activator is the activator itself. This would require the putative activator to have two functionally (but not necessarily physically separate) distinct genetic regulatory elements; one controlled by a growth-factor inducible transcriptional activator such as the fos/jun complex and the second controlled by the transactivator target itself (*Figure 4.4*). Initial activation through the growth-factor inducible transcriptional element leads to growth-factor-independent expression through the second autoregulatory element thus uncoupling gene expression from the signal. There is evidence that such autoregulatory control systems do exist in the context of growth-factor inducible genes, in that the *jun* gene promoter contains a sequence which confers *fos/jun*-dependent transcription upon *jun* itself (8). Although this apparently solves one problem it creates another—the requirement to repress transcription of these genes before completion of the cell cycle to prevent constitutive generation of mitogenic transcription signals and growth-factor independent cell multiplication. This could be overcome by induction of repressors of gene transcription by the prior action of growth-factor inducible activators of gene expression. It is of interest, therefore, that both *fos* and *jun* are both members of a related family of genes, some of which can be demonstrated to antagonize the action of the fos/jun complex (9). It is not clear, however, what role these genes play in the induction of DNA synthesis in response to growth factors.

The final problem lies in the mechanism of linkage between receptor-mediated events, such as activation of PKC or tyrosine kinases, and the activation of gene expression in the nucleus. We have seen in the case of the fos/jun complex that

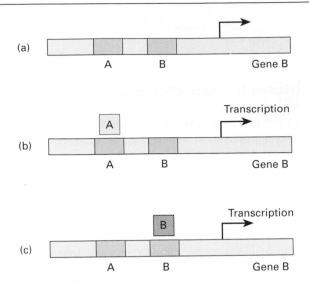

Figure 4.4.   An autoinduction model of growth-factor inducible gene expression. A hypothetical gene (B) has a promoter with binding sites for two transcriptional activator molecules A and B itself. In non-induced cells, neither activator is bound to the promoter of B (a). Upon exposure to growth factors transcription factor A is induced and binds to site A in the B gene promoter (b). The consequent induction of expression of B leads to the formation of functional protein B which binds to site B in the B gene promoter (c) thereby inducing its own transcription. At this point the continued expression of gene B no longer requires transactivator A.

gene activation is set in motion by the prior induction of *fos* gene expression. What activates *fos* gene expression? There must, in other words, exist a mechanism whereby a purely metabolic event (such as protein phosphorylation) occurring at the membrane is translated into the nucleus in the form of gene activation. Again the complete solution to this problem is not wholly clear, but a clear example of such a mechanism can be given. The transcriptional activation factor NF-κB was originally identified as a B-cell specific DNA binding protein that recognizes a *cis*-acting DNA sequence in the regulatory elements of the immunoglobulin light-chain gene. NF-κB has subsequently been found in a variety of non-lymphoid cells and, like many transcriptional regulators, is part of a larger family of related DNA-binding proteins. The absolute levels of NF-κB protein do not change appreciably during the cell cycle but NF-κB-dependent transcription can be induced by treatment of cells with phorbol esters, that is by activation of PKC. It appears that NF-κB in the unstimulated cell is found in the cytoplasm complexed with an inhibitor protein IκB (inhibitor of NF-κB). In the cytoplasmic complex NF-κB has no transcriptional activation activity but can be activated by phosphorylation of IκB which induces breakdown of the complex, permitting NF-κB to translocate to the nucleus and induce transcription (*Figure 4.5*) (10). It is not clear whether IκB is phosphorylated directly by PKC or via

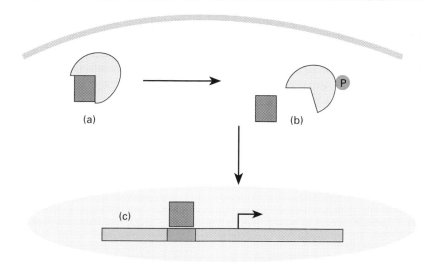

Figure 4.5. The activation of NF-κB activity. The transcription factor NF-κB in (a) is bound to an inhibitor molecule. Phosphorylation of the inhibitor (b) leads to breakdown of the complex and migration of NF-κB to the nucleus where it binds to specific regulatory DNA sequences (c) leading to the activation of gene transcription.

some intermediate PKC-activated serine/threonine kinase. Nevertheless this 'sequestration and release' mechanism, regulated by receptor-stimulated protein phosphorylation, provides at least a working model for the means of translating activation of protein kinases into activation of gene transcription.

It is important to note that the examples described above represent an oversimplification of what is, in many ways, a highly sophisticated and poorly understood system. In particular, many growth factor inducible genes contain different types of regulatory elements (including those conferring cell-type specific regulation) which interact with growth-factor inducible elements in complex ways. In addition, it is clear that post-transcriptional processes, especially the regulation of mRNA stability, appear to play an important role in the regulation of many growth-factor inducible genes. Finally it is also clear that many transcription factors and nuclear regulatory proteins are the targets for growth-factor-activated protein kinase action and the interplay between metabolic events, such as protein phosphorylation, and the regulation of nuclear events, such as gene activation, may prove to be very complex indeed.

7. Unity?

Despite the evident incompleteness of our understanding of the biochemical mechanisms of growth factor action, a broad outline is beginning to emerge. Growth factors interact with specific cell surface receptors, which leads to the rapid activation of signalling systems, often involving protein phosphorylation. These rapid signalling events are translated into regulation of gene expression, which appears to involve a cascade of time-dependent processes, eventually culminating in activation of the processes leading to the entry into DNA synthesis, and execution of the remainder of the cell cycle. At each stage signalling events diverge, resulting in a broad spectrum of cellular responses, a subset of which involve the 'core machinery' of cell-cycle control.

8. Further reading

8.1 Review

Lewis, B. (1991). Oncogenic conversion by regulatory changes in transcription factors. *Cell* **64**, 303–12.

9. References

9.1 Gene expression and the mitogenic response

1. Smith, J. C. and Stiles, C. D. (1981). Cytoplasmic transfer of the mitogenic response to platelet-derived growth factor. *Proc. Nat. Acad. Sci. USA* **78**, 4363–7.

9.2 Induction of gene expression by growth factors

2. Almendral, J. M., Sommer, D., Macdonald-Bravo, H., Burckhardt, J., Perera, J., and Bravo, R. (1988). Complexity of the early genetic response to growth factors in mouse fibroblasts. *Mol. Cell Biol.* **8**, 2140–8.
3. Greenberg, M. E. and Ziff, E. B. (1984). Stimulation of 3T3 cells induces transcription of the *c-fos* proto-oncogene. *Nature* **311**, 433–8.

9.3 The TPA response element

4. Angel, P., Imagawa, M., Chiu, R., Stein, B., Imbra, R. J., Rahmsdorf, H. J., Jonat, C., Herrlich, P., and Karin, M. (1987). Phorbol ester-inducible genes contain a common *cis* element recognized by a TPA-modulated *trans*-acting factor. *Cell* **49**, 729–39.

9.4 fos and jun

5. Curran, T. and Franza, R. (1989). Fox and jun; the AP-1 connection. *Cell* **55**, 395–7.
6. Rauscher, F. J., Cohen, D. R., Curran, T., Bos, T. J., Vogt, P. K., Bohmann, D., Tijian, R., and Franza, B. R. Jr (1988). Fos-associated protein p39 is the product of the *jun* proto-oncogene. *Science* **240**, 1010–16.
7. Koudarides, T. and Ziff, E. (1988). The role of the leucine zipper in the fos-jun interaction. *Nature* **336**, 646–51.

9.5 jun auto-regulation and inhibition

8. Angel, P., Hattori, K., Smeal, T., and Karin, M. (1988). The *jun* proto-oncogene is positively autoregulated by its product, Jun/AP-1. *Cell* **55**, 875–85.

9. Chiu, R., Angel, P., and Karin, M. (1991). Jun-B differs in its biological properties from, and is a negative regulator of, c-Jun. *Cell* **59**, 979–86.

9.6 NF-κB and nuclear translocation

10. Nolan, G.P., Ghosh, S., Liou, H.C., Tempst, P., and Baltimore, D. (1991). DNA binding and I kappa B inhibition of the cloned p65 subunit of NF-kappa B, a rel-related polypeptide. *Cell* **64**, 961–9.

Glossary

Affinity: A measure of the strength of interaction between two molecules.

Autocrine: A system in which a signal is made by, and acts upon, the same cell.

Autophosphorylation: The addition of phosphate groups by an enzyme to an identical enzyme.

cDNA: The DNA copy of an mRNA molecule.

Domain: A discrete structural feature of a protein.

Endocrine: A system in which a signal produced by a cell acts upon an anatomically remote target cell.

Endothelial cells: The cells which line the walls of blood vessels.

Erythroblast: A cell which gives rise to mature red blood cells.

Establishment: The process by which cells acquire an infinite proliferative potential.

Extracellular: Located outside the plasma membrane of a cell.

G protein: A protein which binds guanidine nucleotides.

G1: The period of the cell cycle which precedes S phase.

G2: The period of the cell cycle between the completion of S phase and the onset of mitosis.

Gel electrophoresis: A technique in which proteins are physically separated on the basis of their mass as the result of migration in an electric field through a porous matrix.

Glycosidase: An enzyme which cleaves the bonds between sugar residues in carbohydrates.

Haemopoiesis: The cellular processes which give rise to the cells which are found in blood.

Heterodimer: A molecule produced by association of two different proteins.

Homodimer: A molecule produced by association of two identical proteins.

Intracellular: Located inside the plasma membrane of a cell.

Ionophore: A chemical which permits small charged molecules to pass across the plasma membrane.

Kinase: An enzyme which catalyses the covalent addition of phosphate residues to substrates.

Ligand: A molecule which binds in a specific manner to a receptor.

Metalloproteinase: An enzyme which degrades proteins and employs a bound metal ion for catalysis.

Micronutrient: A low molecular weight chemical compound required for cell viability.

Mitogen: A molecule which induces cell multiplication.

Mitosis: The phase of the cell cycle which involves nuclear division and the equal partitioning of chromosomes into new cells.

Neuropeptide: A small protein with intercellular signalling functions in the nervous system.

Paracrine: A system in which a signal produced by a cell acts upon another cell in close physical proximity.

Phospholipase: An enzyme which cleaves phospholipids.

Plasma: The liquid component of blood.

Protease: An enzyme which degrades proteins.

Quiescence: A state in which cells are not actively dividing but retain the capacity to proliferate.

Receptor: A molecule which binds a ligand and, as a result, exhibits a change in function.

Repressor: A molecule which inhibits a biochemical process.

Retrovirus: A virus whose life cycle involves the production of a cDNA copy of an RNA genome of the virus particle and integration of the cDNA into the genome of the infected cell.

S phase: The phase of the cell cycle in which DNA synthesis occurs.

Senescence: A state in which a cell has permanently lost the ability to proliferate.

Serum: The liquid fraction of clotted blood.

Signal sequence: A polypeptide sequence within a protein which directs synthesis on ribosomes associated with endoplasmic reticulum. Proteins with signal sequences are destined for export from the cell or association with membranes.

Stoichiometry: The proportion of a group of otherwise identical molecules which have undergone some specified alteration.

Substrate: A molecule which is the target for the action of an enzyme.

Transcription: The process by which an RNA copy is made of a DNA sequence.

Transfection: The introduction of exogenous DNA molecules into cells by physical or chemical techniques.

Transformation: A genetic alteration to a cell which renders it able to grow without the requirement for exogenous mitogens. Transformed cells are often able to form tumours *in vivo*.

Index